Data Analytics in Sensing Devices

Data Analytics in Sensing Devices

Dr. Ambika N

CWP

Central West Publishing

Disclaimer
Every effort has been made by the publisher, editors and authors while preparing this book, however, no warranties are made regarding the accuracy and completeness of the content. The publisher, editors and authors disclaim without any limitation all warranties as well as any implied warranties about sales, along with fitness of the content for a particular purpose. Citation of any website and other information sources does not mean any endorsement from the publisher, editors and authors. For ascertaining the suitability of the contents contained herein for a particular lab or commercial use, consultation with the subject expert is needed. In addition, while using the information and methods contained herein, the practitioners and researchers need to be mindful for their own safety, along with the safety of others, including the professional parties and premises for whom they have professional responsibility. To the fullest extent of law, the publisher, editors and authors are not liable in all circumstances (special, incidental, and consequential) for any injury and/or damage to persons and property, along with any potential loss of profit and other commercial damages due to the use of any methods, products, guidelines, procedures contained in the material herein.

A catalogue record for this book is available from the National Library of Australia

ISBN (print): 978-1-922617-35-4

Preface

Devices like sensors, Internet-of-things (IoT), and Radio frequency Identification (RFID) generate enormous amounts of data. The machines are a part of several environments. The information is collected forms a large database. The collection investigates to gain better insight into the current status of the sensed entity or environment. The knowledge set aims to improve the system. It helps to obtain good profit and makes future predictions. Several metrics are used to conduct the examination.

The book discusses the technology architecture, usage, and evaluation using various metrics. The different algorithms used in the functioning of the contribution are summarized. The writing also suggests the future scope of the work in diverse applications.

About the Author

Dr. Ambika is a MCA, MPhil, Ph.D. in computer science. She completed her Ph.D. from Bharathiar university in the year 2015. She has 16 years of teaching experience and presently working for St. Francis College, Bangalore. She has guided BCA, MCA and M.Tech. students in their projects. Her expertise includes wireless sensor network, Internet of things, cybersecurity. She gives guest lectures in her expertise. She is a reviewer of books, conferences (national/international), encyclopaedia and journals. She is advisory committee member of some conferences. She has many publications in National & international conferences, international books, national and international journals and encyclopaedias. She has some patent publications (National) in computer science division. She is presently working for St. Francis college, Bangalore.

Foreword

Information Assurance (IA) is a combination of technologies and processes that are used to manage information-related risks. IA is not just about computer security, the protection of data in storage or while it is being processed; it is also about the protection of data in transit. IA is a composite field involving computer science, mathematics, database and network management, user training, and policy issues. A common objective of work on IA in these fields is to protect and defend information and information systems by ensuring their availability, integrity, authentication, confidentiality, and non-repudiation so that the right people can access the right information at the right time.

The Internet, social media, smartphones and tablet computers have been playing a larger role in our daily lives. The majority of computers, whether in large corporations, in small businesses, or at home, are connected together in a network that creates a global community. People have become increasingly dependent on computer networks in many aspects of their lives — from communication, entertainment and financial transactions to education and government services. Most people understand that global economic infrastructure is becoming increasingly dependent upon information technology, and no information system is 100% secure. Information security is one of the topics that everyone knows of, but most are not really aware of the finer details. Many computer users simply think that their firewall and antivirus software provide them with all the protection they need to keep their computers secure. However, as malicious hackers become more resourceful, and users add more and more information into a growing number of databases, there exists an increased exposure to hacker attacks, information espionage, and other security breaches. Information systems—operated by governments and commercial organizations—are vulnerable to attacks and misuse through their Internet connections. Workstations connected to the Internet are currently the most common targets of malicious hackers. As a result, information assurance is a very serious concern for individuals, businesses, and governments. Not only do we need to be aware of how attacks are perpetrated, but we also need to learn how the systems can be protected against different attacks.

This book provides a valuable window on information assurance and covers the necessary components from detecting Internet worms distributed via e-mail to securing mobile communication devices. Firewalls are a critical technology to control incoming and outgoing network traffic, thereby blocking unwanted traffic and suspicious connections. They must be configured with a set of filtering rules and, like any software application, must be constantly patched to address new vulnerabilities. Authentication verifies the identity of each user or examines the validity of a device. Currently, passwords are the most commonly used authentication scheme. Because of its uniqueness, biometrics, such as fingerprint, iris or facial images, are becoming a promising means of authentication. Bill Gates predicted that biometric technologies will be one of "the most important IT innovations of the next several years" at a Gartner Group IT/Expo event held in 1997. In order to ensure the current user is the same person that logged onto the system, research efforts have been devoted to continuously verify the user's identity using biometrics. Security and privacy are closely related. When developing an information security solution, we need to consider its impact on privacy and combine security risk assessment techniques with privacy risk assessment techniques. Risk assessment is a critical process to define both the probability and impact of undesired events. Its objective is not to eliminate the risk, but to provide the policy and methodology by which risks could be managed.

The challenges in information assurance are both difficult and interesting. People are working on them with enthusiasm, tenacity, and dedication to develop new methods of analysis and provide new solutions to keep up with the ever-changing threats. In this new age of global interconnectivity and interdependence, it is necessary to provide security practitioners, both professionals and students, with state-of-the art knowledge on the frontiers in information assurance. This book is a good step in that direction.

I am delighted to write this foreword, not only because Dr. Ambika has been associated with me since 2016 and colleague for more than eight years, but also because I believe deeply in the educative value of interpretive discussion for all students, especially in a democratic society. I also believe that teachers at every level and stage of their career can enrich and strengthen their teaching by learning the discussion leading recent topics like IoT, Data Analytics and practices presented in this book. Participating in interpretive discussions can help teachers and students alike learn to use their minds with power and pleasure.

This isn't just another book about data analytics. It's a book about detailing of data analytics with practice orientations: the very heart and soul of good writing. Dr. Ambika, has majorly focused on real-life aspects of data analytics. She also focused on theory to understand the basic concept of data analytics. She uses several use cases to explain the real-life usage of the same. Furthermore, she explained the applications, challenges and proposed solutions in data analytics. You can bet that any genuinely thoughtful book on data analytics. When students feel that kind of power, data analytics is no longer a mindless chore; it's wonderful thoughtful concepts —and the options are exciting.

Dr. Mayank Singh
Senior Scientist
Consilio Research Lab, Estonia

Recognising the rapid advance in the field of Information Technology in recent years, there is a need to bring in latest development into the book. Accordingly, this book is written with latest developments in the field of IOT, RFID.   The sensing instruments like the Internet of Things (IoT), Sensors, and RFID collect large datasets. This dataset undergoes various kinds of analysis to understand the system and forecast. This methodology has helped the growth of technology aiding humans in day-to-day life. The book is a summary of the different kinds of technology. The first chapter gives an introduction to Data analytics. It introduces the reader to various algorithms and their usage in sensing devices. The second chapter details the architecture of IoT instruments. It discussed different sorts of data analysis with its usage in various applications. The third chapter explains sensor architecture. It discussed diverse applications supported by the sensor world and diverse algorithms. The fourth chapter details the architecture of RFID, the characteristic of the technology, and the usage of analytics algorithms in the technology. All the chapters highlight the challenges faced by the technology and its future scope. I am sure the book in the present form will be very useful to all readers.

Dr. R. N. Subba Rao
Principal
St. Francis college, Koramangala

The Internet of Things is a future technology in the present. It is trying to automate and facilitate our daily life interconnecting Smart Objects and objects (sensors and actuators) between themselves. With all this data, it is creating a smarter network to make decisions and facilitate these automations. However, all the data which is travelling between the objects and the Internet is terrific huge. This is why we have a lot of objects in our life: Smart TVs, smartphone, Smart Cars, microcontrollers, Smart Bands, Smart Tags (RFID and NFC), and more. Besides, we have to add to all this data our data too. This last one is on the Internet and is generated in an automated way or from us like in Online Social Networks and other applications that we have used or other uses to facilitate all the things like in the government.
Then, we need to analyse this data to facilitate the use of it to people and objects, and to create automations and useful statistics. To achieve this, we have to prepare the data to read it. This implies to clear it, read it, and know from where it is because probably the work depends on the object (format, type, frequency), and keep it to analyse it. Maybe to search trends and topics, create information, classify it, and/or use it. Therefore, data analytics is so important.

One example of it is the Industrial Internet of Things or Industry 4.0, where the Internet of Things was born. Kevin Ashton proposed to use RFID tags in supply chains to obtain in an automated way all the information using RFID readers and computers. In this way, it can manage in a semiautomatic or automatic way where each part is, take the information like the time and the stock, and facilitate to the final user all the traceability and stock system. With data analytics, we can programme alarms about the stock, know who need more or less items, who is a better client, and improve deliveries.

This is the reason why data analytics and the Internet of Things are so close together.

Cristian González García
Departamento de informática / Computer Science Department

Facultad de ciencias, planta 3, despacho 197, Oviedo, Asturias (Spain)
Universidad de Oviedo

The book titled "Data Analytics in Sensing devices" has beautifully brought up the analysis done with the reading collected by IoT, sensors, and RFID machines. The Sensing instruments like the Internet of Things (IoT), Sensors, and RFID collect large datasets. This dataset undergoes various kinds of analysis to understand the system and forecast. This methodology has helped the growth of technology aiding humans in day-to-day life. The book summarizes the different kinds of technology. It introduces the reader to various algorithms and their usage in sensing devices. In the following chapter's we see architecture of IOT Instruments discussed in detail with its various applications, the diverse applications of sensor architecture supported by the sensor world and its algorithms, the architecture of RFID, its characteristic of the technology and the usage of analytics algorithms in the technology. Overall, the book highlights the challenges faced by the technology and its future scope.

Dr. Baswaraj Biradar
PRINCIPAL
HKBK Degree College, Bangalore

Data Analytics aims in using better tools and equipment to analyse big datasets and take appropriate decisions and make strategies. The book addresses the solutions given to problems related to the business by collecting data from sensors, IoT devices, and RFID machines. The ways of exploration of concepts in this book is good.

Dr. Mansaf Alames
Editor-in-Chief, Journal of Applied Information Science
&
Associate Professor & Young Faculty Research Fellow, DietY, Govt. of India
Department of Computer Science
Jamia Millia Islamia, New Delhi

Data analytics is an expansive term that incorporates numerous different kinds of data examination. Any sort of data can be exposed to data analytics methods to get knowledge that can be utilized to further develop things. Data analytics strategies can uncover patterns and measurements that would somehow, or another be lost in the mass of data. This data can then be utilized to enhance cycles to expand the general proficiency of a business or framework.

IoT reception is becoming sought after and regularly utilized by business moguls. In any case, since the quantity of associated gadgets will in general expand, associations think that it is difficult to deal with the immense measures of created data. Since the data assembled from IoT are less organized, the data sets are very many-sided. IoT Analytics devices assist associations with utilizing complex IoT datasets, increment profit, further develop items, and improve items considering the requirements of their clients. IoT and data resemble twins that are essentially connected. At the point when organizations get hold of indispensable data to dissect, they use the data gathered by IoT gadgets. This is the place where the association of IoT and data analytics happens.

The blend of data analytics and IoT sensors likewise assists organizations with deciding when gadgets need support by estimating hotness, vibration, and other critical figures. Brilliant gadgets can likewise send data to administrators about wear, expected breakdowns, and even conveyance plans. This doesn't just work with standard gadget upkeep yet additionally sponsors logical support. This book presents the detailed elaboration about the data analytics in its various operations including the various approaches used in data analytics, algorithms, IoT sensor data analysis, RFID, etc.

Jyotir Moy Chatterjee
Assistant Professor (IT)
Ambassador (Scientific Research Group in Egypt, Bentham Science Publishers & Sustainable Procurement & Supply Chain)
Lord Buddha Education Foundation (Asia Pacific University of Technology & Innovation)
Kathmandu, Nepal-44600

Data Science is one of the emerging areas. It is not just useful for computer science field rather useful for all fields. Data science widely talks about the prediction and forecast of all aspects of the business. Where, prediction is used to identifying the certainty, forecast represent the range of certainty. Data science provides various techniques for understanding the data, prediction and forecast.

Data collection, cleaning, designing and analysis are the fundamental functions of data science. Interpretation of numerical data for the purpose of pattern recognition helps in making various business decisions. All most all businesses decisions have become increasingly dependent on data science in many aspects and getting helping in the growth prominently.

The author of the book "Data Analytics in Sensing Devices ", Dr. Ambika. N, is an expert and her experience is visible in the form of this book. The book is focussing on fundamental as well as in-depth knowledge of data science. This book discusses about various predictive algorithms, analytics model and Learning analytics. IoT data analytics is explained beautifully. This book is discussing about various use cases including i) Smart metering ii) Smart transportation iii) Smart supply chains iv) Smart agriculture v) Smart grid vi) Smart traffic light system.

The book focuses on the analysis done with the reading collected by IoT, sensors, and RFID machines. This book provides a valuable window on data science and predictive analysis and covers necessary steps and techniques in the right direction. IoT architecture, Sensor architecture, and RFID architecture for data analytics, all are elaborated with various Applications and challenges to adopt the technology is explained flawlessly. She is sharing real life scenarios with the readers.

I congratulate Dr. Ambika. N for bulging her experience in the form of this book and wish her all the best.

Dr. Kavita Saini
Professor
Galgotias University, Delhi NCR, India

Data Analytics in Sensing Devices is an amazing book in today's time, where data analytics applications in various industries is growing and the usage of sensing devices is increasing fast and continues to grow.

This book authored by Dr. Ambika Nagaraj explains the concept of data analytics briefly. It will prove to be very helpful in understanding the data analytics concept. Along with it, the role data analytics can play in analyzing different readings obtained by using sensing devices like sensors, etc. is also explained in this book.

The book has kept a great focus on the analysis done with the reading collected by using sensors, RFID device, etc. It consists of various examples of different applications and use cases book to explain the different aspects of the topic. The book has been written in such a way that it covers the various different aspects of the topic very well with a systematic presentation.

Prakash J
PSG College of Technology
Coimbatore, India

Table of Contents

1

INTRODUCTION TO DATA ANALYTICS

Abstract

Data analytics is a computation methodology where the collected unprocessed information undergoes examination to conclude. The procedure uses many robotic algorithms to understand the collection. The business applications optimize their performance doing the investigation. The chapter is an introduction to the topic. Different investigation procedures are discussed. The application of the methodology in diverse domains is summarized in the chapter.

1.1 Introduction

The business community (North, 1997) is growing every day and bringing new products. But sometimes this advancement fails if the introduced product does not have any market value. Hence analysis is a big part of any business setup. Various investigations w.r.t to the dependent parameters are to be analyzed. The outcome will give a clear picture of the new business setup. The data examination (Runkler T. A., 2016) (Kambatla, Kollias, Kumar, & Grama, 2014) (Tsai, Lai, Chao, & Vasilakos, 2015) is enhanced investigation methodologies that operate on big information datasets. The scheme identifies the changes and its procedure to react to the changes. The three bulky knowledge are quantity (Awan, Brorsson, Vlassov, & Ayguade, 2015), type, and rapidity. Many enterprise possibilities embrace Overcoming bankruptcy and rehabilitation.

The sophisticated inquiry is the higher means to determine the new client sector, identify the most suitable trader, correlate goods of resemblance, and recognize auction seasonality. It is an assortment of interrelated practices. It includes future analytics, facts mining, numerical scrutiny, and composite SQL. The catalog can wrap records revelation, mock intellect, lingo dispensation, and folder potential that maintain analytics. The customer is characteristically a trade forecaster who tries to notice new commerce particulars that no one in the venture knew before. The market analyst needs bulky data with an ample amount of details. Detection analytics permits diverse kinds of systematic methods. It includes SQL inquiry, statistics mining, numerical investigation, etc. The information aggregated in knowledge repositories fluctuates from learning gathered for scrutiny. A divergent variety of examinations may have various information sets. Some diagnostic methodologies direct an industry forecaster to make ad hoc methodical facts set per reasoned scheme. The latest tapping of these resources for analytics means that so-called ordered statistics is formless information (Yafooz, Abidin, Omar, & Idrus, 2013) and semi-structured dataset (Abiteboul, 1997).

Data Analytics starts from the mining procedure. Information Extraction (Goebel & Gruenwald, 1999) is a computerized or semi-mechanical technique for pulling facts, regulations, and practices from enormous magnitudes of datasets. They have two classes. The administered education (Muhammad & Yan, 2015) creates a prototype that illustrates the connections between a destination parameter and visionaries by examining an exercise collection in the learned dataset. A terminus variant is an outcome content estimated. The forecasters are autonomous variants impacting the objective parameter. It has two groups. Arrangement (Berkhin, 2006) assesses the input to obtain an unconditional destination outcome. The purpose of reversal is to calculate a mathematical terminus variant. Unverified knowledge (Khanum, Mahboob, Imtiaz, Ghafoor, & Sehar, 2015) comprehends and defines the design of an information collection. It includes grouping investigation and linkage examination. The purpose of collection research is to discover clusters of documents matching particular attribute sets and different from other collections. The connection investigation (Kumar & Chezian, 2012) pulls fascinating collaboration controls from a routing details cluster. Figure 1.1 portrays the steps in Data analytics.

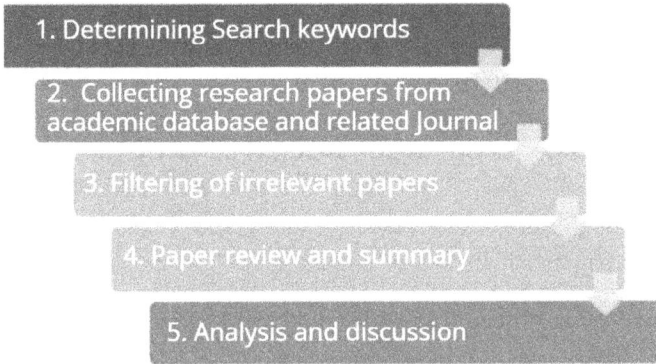

Figure 1.1 Adopted Procedure in Data analytics (Park, Lee, Park, & Kim, 2021).

1.2 Predictive algorithms

Forecast procedures (Vilalta, Apte, Hellerstein, Ma, & Weiss, 2002) perform a critical function in operations administration. The capability is to foretell service queries in processor systems. It answers those signals by implementing restorative operations, brings various advantages. It detects framework collapses on a few hubs. It can limit the range of those crashes to the complete arrangement. The forecast guarantees the constant requirement of organization assistance through the computerized implementation of repair operations. Figure 1.2 represents the architecture of air quality monitoring and prediction system.

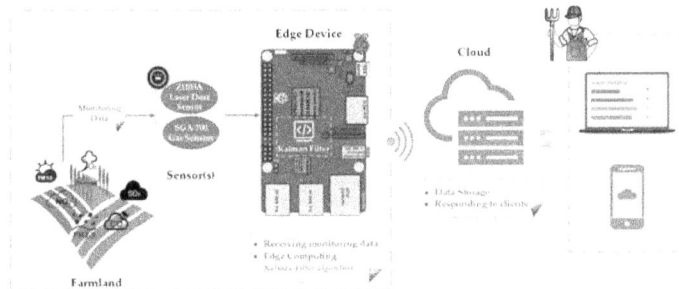

Figure 1.2 The architecture of air quality monitoring and prediction system. (Lai, Yang, Wang, & Chen, 2019).

Using archival data, one can make forecasts. Some of the parameters predictable cover the memory or disk usage on an owner or collection of managers. Other variables include the quantity of Hypertext Transfer algorithm transactions per moment in a Web host and the state of a system device at a given time. The forecast is scanning for a method resembling the features of the query. Critical constituents are the discrete or constant characteristics of the info. They can be checked at regular period or not. The knowledge collection over intervals resembles immediate preferences. The sampling tool depends on the information assemblage process. It is either cyclic or triggers sampling. Learning accumulated by recurrent data includes administration measures such as usage and end-to-end acknowledgment point to a probing assistant.

Choosing the appropriate forecast procedure varies with two constituents. The type of prediction divides into short-term and long-term predictions. It ascribes a definite interest to both courses. In machine operations, it is logical to illusions short-term forecasts on the scale of hours to minutes. Long-term foresight is the scale of days, weeks, or months. The next part can be either numeric or Boolean.

1.2.1 Random forest model

Irregular Forest procedure (Shaik & Srinivasan, 2019) is an Administered Grouping method. It analyzes the results by building several Classifiers intending to accomplish the correct forecast. It applies to sample information where the blocks get created. The outcome individuals couples to foretell the course description. The organization of a massive quantity of knowledge guides has limited precision in the outcome. It is a part of many purposes where a tremendous volume of learning is incorporated with the Decision Tree Distribution methodology. It is easy to understand for processor specialists and clients without arithmetical knowledge. It does not need any hybrid confirmation. It uses Adaboost and Bootstrapping procedures to build various classifiers. Figure 1.3 portrays the architecture of intrusion detection system.

```
┌──────────────────────────────────────────────────────────┐
│ Data detection center                                      │
│  ┌─────────────────────┐        ┌─────────────────────┐   │
│  │ Establish a random  │───┐    │ Classification results│  │
│  │ forest classifier   │   │    └─────────────────────┘   │
│  │ module              │   │              ⬆                │
│  └─────────────────────┘   │    ┌─────────────────────┐   │
│           ⬆                │───▶│ Test random forest  │    │
│  ┌─────────────────────┐   │    │ classifier          │    │
│  │ SMOTE               │──┘     └─────────────────────┘    │
│  └─────────────────────┘              ⬆                    │
│           ⬆                                                │
│  ┌─────────────────────┐        ┌─────────────────────┐    │
│  │ Training set        │        │ Testing set         │    │
│  └─────────────────────┘        └─────────────────────┘    │
└──────────────────────────────────────────────────────────┘
                          ⬆
┌──────────────────────────────────────────────────────────┐
│ Data perception center                                     │
│     ┌──────────────────────────────────────────────┐      │
│     │ The dataset of wireless sensor networks       │      │
│     └──────────────────────────────────────────────┘      │
└──────────────────────────────────────────────────────────┘
```

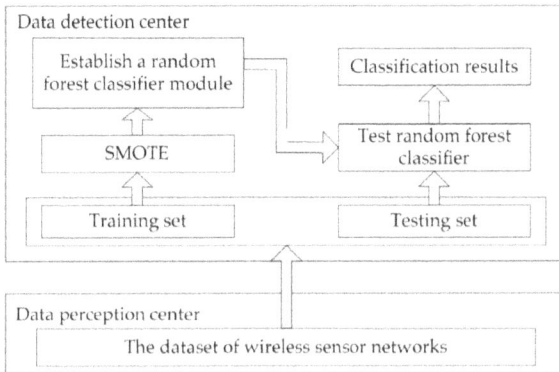

Figure 1.3 The architecture of intrusion detection system (Tan, et al., 2019).

The system (Tan, et al., 2019) divides the outcome into a (routine) dataset and four kinds of invasion. The work employs SMOTE to make the database. The system rebuilds the workout collection. The operation optimizes the framework. The suggestion uses the Random Forest procedure to teach the database. The recommendation uses a classifier to observe the attack pattern. The work uses KDD Cup 99 database. It has a workout and trial collection. The aggregation is a system gridlock cluster designed by the MIT Lincoln workshop by mimicking the regional location atmosphere of the U.S. Air Force. The database has 41 types of features. The work has five categories. Four invasion classes considered are Denial of Service attack, user to root invasion, Probing, root to local attack. The investigation executes using a 2.6 GHz Intel Core i5-3320M system with 4GB RAM.

The suggested procedure (Mohnen, Rotteveel, Doornbos, & Polder, 2020) is the answer to foretell a person's health cost. The system accesses non-public dataset. This information links to personality and vicinity stage. It contains the complete Dutch populace. Anomaly dataset is investigated in a protective isolated surrounding. Locality is operational using the vicinity identity. It is a tiny and more accurate procedure of 4-numeric postal regulations. The framework uses metropolitan catalog information counting house location, rearrangement appointment, and social demographic features. The dataset has amount of domestic and social financial grade. The well-being knowledge centre aggregates and administers wellbeing of all

Dutch fitness assurance enterprises sheltered by the physical condition cover rules. It includes the expenses of obligatory co-
expenditure and deduction, apart from other out-of-compartment
expenses. The state fitness concern organization creates a yearly
summary of prearranged prescription per resident. It is based on
Anatomical healing compound categorization. The datasets includes
ATC identification on 4-numeric stages. The authors have used
21,559,510 datasets. The dependent variant is medium over a pre-
set epoch of the yearly personality standard wellbeing expenditure.
It includes all overheads that were enclosed by the essential fitness
assurance in the Netherlands.

The frame (Kaur, Kumar, & Kumar, 2019) is a judgment assistance
operation. It operates in the cloud and computer knowledge procedures. It checks their enforcement. Victims reach from those implemented with quickly obtained and affordable sensing devices at
their houses to the located at unknown positions or a reach from the
pharmaceutical provisioning. The assistance seekers include all the
sufferers attending the dispensaries or labs who do not have the vis-
ibility of the specialists. They are outfitted with essential therapeutic
tools and assist workers acting as the intermediate. They dispose of
the wellness knowledge gathered from victims using the detector
arrangements onto the cloud for the physicians to reach and answer. All the accumulated fitness care learning is shifted initially
into the portable tools through a detector arrangement. The device
system may be Bluetooth, Wi-Fi utilization, or USB-based association. Mobile devices act as IoT operators and are employed to communicate the wellness knowledge onto the cloud. K-Nearest Neigh-
bor procedure is most adjacent acquaintance information limits. It
detects the anonymous knowledge scores. It analyzes the evidence
features according to the polling operation. The system is proficient
to a paired arrangement and multi-class puzzles. The SVM method
generates large hyper-planes in dimensional locations. It maximizes
the detachment among score features, and backer vectors are employed to build a hyperplane. The SVM presents more reliable exactness. Environment for Knowledge Analysis (WEKA) is an open-
source tool. The dermatology database includes 366 specimens and
six levels. It has the highest efficiency of 97.26%.

The system (Javeed, Jalal, & Kim, 2021) is healthcare observed practice. It has three stages. In the pre-processing and segmentation

stage, the information of all three varieties of detectors has been refined using median, band-pass, and rolling medium filters. Denoised knowledge splits into overhanging shutters according to the measurement of each stage of bodily movements. Event area properties and analytical peculiarities are done in the characteristic extraction period. In symbolization with a group stage, component vectors assign using a caught random forest procedure. The recommended design has sound replacement and preprocessing measures to attain error-free knowledge. The sonance of many inertia's sensing beacons uses a median filter for IMU detector, band-pass filter for EMG, and rolling medium filter for MMG. After eliminating the sound, panes for all sign types have been included based on 5 seconds. Time-domain characteristics and statistical peculiarities such as zero exchanges incline flag variations, skewness, and peak-to-peak characteristics maintain the ordering means. Then, Linde-Buzo-Gray (LBG) algorithm has signifying components. These characteristic vectors are into a haphazard growth procedure for organization and exercise identification for wellness.

1.2.2 Generalized linear model

Generalized linear standards are employed to examine associations between parameters and a reaction. They have recognized connections and answers are dispersed. It permits the volume of the difference of each dimension to be a process of its forecasted content.

Linear standards embody both routine and arbitrary associates. The mistakes consider having regular allotments. The member's logical approach is a least-squares hypothesis. It is in its traditional structure. It has one misconception element. The attachments for numerous falsehoods develop investigation in organized experimentations and survey information. Methods created for non-normal knowledge comprise probated estimation. The binomial variant has a parameter correlated to an unstated underlying forbearance allocation and contingency data collection. The allotment is multinomial and the recurring function of the prototype. The process (Nelder & Wedderburn, 1972) is recurrent weighted linear degeneration. It receives the highest probability calculations of the parameters with compliances allocated. It uses exponential lineage and periodic results that can be made linear by an appropriate conversion. The system has a category of generalized linear standards. It has a cooperative process for serving them based on possibility. The architecture combines the frequent and unexpected elements in the prototype to create the generalized linear measure. The key of the highest probability equations is identical to a recursive weighted least-squares methodology with a significant operation.

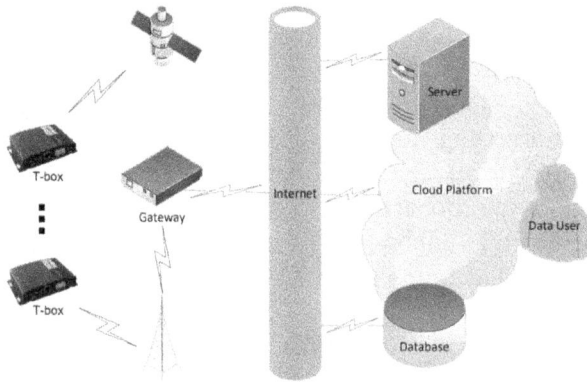

Figure 1.4 Schematic diagram of data acquisition, transmission, storage and application (Sun, Bi, Guillen, & Pérez-Marín, 2020).

The hardware (Sun, Bi, Guillen, & Pérez-Marín, 2020) has Raspberry Pi, the detector grid, and the Wi-Fi model. The programming section has stockpile and consumer design. The system observes the intensity of air contaminants such as SO_2, NO_2, CO, O_3, PM2.5, and PM10. The system completes the routine collection of data. The work ships the different impurity information to the hardware. Raspberry Pi executes the Kalman Strainer procedure. The system estimates anticipated contents using the sampled data. The hardware transmits the details to the warehouse using the Wi-Fi framework. The repository caches the knowledge. It shares it with the consumer and demonstrates the client with facts. The work is an analysis conducted on IoV knowledge assistance in China. It is a high-tech corporation delivering intellectual grid benefits. It assembles details from their wise web automobiles. The automobile has been pre-configured with a telematics package. It contains a GPS detector, a motorcar situation device, and an unconnected communication division. The system updates the dataset (from the time the vehicle has started till it has stopped). It assembles data at the machine tier to decrease facts dispatch and warehouse prices. T-package disseminates with the dataset every 30 seconds. The panel communicates after 30 minutes (when the machine shuts down). The details from the warehouse possess distinctive motorcar labels, GPS orbit knowledge, and automobile state details. The same is represented in Figure 1.4.

This architecture (Eletter, Yasmin, Elrefae, Aliter, & & Elrefae, 2020) communicates medical indicators such as heart pace, blood stress, and weight. A professional at the fitness maintenance establishment will examine the knowledge to deliver appropriate therapy. The database has clinical documents of 299 HF of the sick. It includes medical and routine learning. The knowledge base has 13 variants. It arbitrarily separates into 70% or 209 topics for exercise and 30% or 90 subjects for testing. The activity knowledge teaches five device understanding procedures to create classifiers. It classifies sick into dead patients and healthy patients. The outcome identifies the framework with the peak categorized appropriateness.

1.2.3 Gradient boosted model

Gradient growing procedure (Liu & Pan, 2021) is a collaborative education methodology. The category and degeneration issues use the algorithm. It can deliver an adequate prototype having vulnerable trainees. It creates and simplifies the group standard in a stage-wise style by elevating an unplanned loss procedure. It forms its instance from the earlier loss operation of an opposing gradient in a repetitive mode.

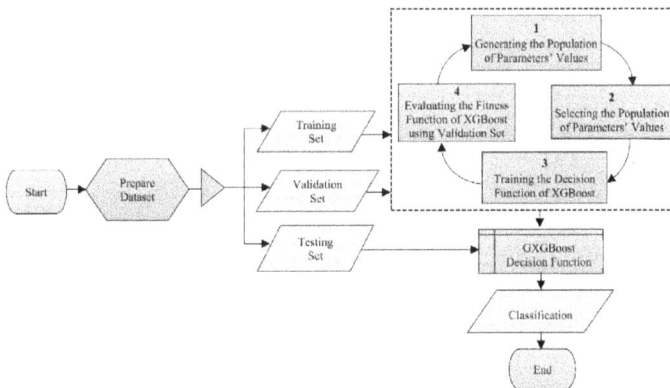

Figure 1.5 Flowchart of proposed genetic-based extreme gradient boosting (GXGBoot) Model. (Alqahtani, Gumaei, Mathkour, & Maher Ben Ismail, 2019).

The genetic procedure (Alqahtani, Gumaei, Mathkour, & Maher Ben Ismail, 2019) yields arbitrary text. XGBoot classifier uses the outcome to generate a new finding border with the tallest hereditary wellness. The system has four phases - developing the inhabitants of features, choosing the appropriate ones, teaching the conclusion procedure, and assessing the healthiness operation. The system examines the database by simulating using the same. The work applies appliance education strategies for witnessing and categorizing Rejection of Service invasions. The detector catches invasions design from the regular gridlock. The imitation information includes 23 characteristics mined utilizing LEACH directing algorithm. Figure 1.5 depicts the same.

The isolated perception information and Gradient Augmented degeneration computed crop in the Kharif period (Arumugam, Chemura, Schauberger, & Gornott, 2021). The time duration assumed was 2003 to 2017, and the area considered was 500 m spatial resolution in India. The authors created and evaluated individual standards for every phase. The system availed Area Index values at 500 m spatial resolution. The time interval was for eight days. The resampling considered leaf index and harvest into consideration. The work utilized an area of 5 km to assess the produce. The system found the mean of the index contexts. The work assigns district-level products value to all the grid cells. The middle product measurement per precinct in India is around 75,000 hectares. It has a total produce harvest size of 44 million HA. The system taught GBR examples per condition. It utilizes 5 km resolution and uses the outcome on 500 m MODIS LAI area resolution to forecast crop outputs at 500 m resolution.

The suggested IoT-based detector (Alfian, et al., 2020) has two machines. Raspberry Pi is an inexpensive SBC of 85.60 mm × 53.98 mm × 17 mm. It weighs approximately 45 g. It is competent in managing different intake and result processes. The Sense-HAT panel assembles atmospheric states. Python-based software collects detector readings from Sense-HAT. The IoT-based machine constantly accumulated warmth and wetness information and dispatched it to stockpile. The trailing method permits various logistics associates to develop diverse yield details before shipping to EPCIS. The manufacturer's stock stocks recent products. It uses an RFID gateway. The outcome details ship to the EPCIS. The handheld reader document yields facts. The transporter supplies the derivative to the client. They read the outcome data utilizing a handheld reader. The developed knowledge ships to the EPCIS using transporter mobile.

1.2.4 Prophet algorithm

The prophet is a method for foretelling time sequence information based on an additive standard. The non-linear movements are suitable with annual, weekly, and everyday seasonality. It performs satisfactorily with period succession that has substantial seasonal consequences and many seasons of documented knowledge. Prophet handles missing information and changes in the tendency and typically endures outliers well.

The detector (Parise, Manso-Callejo, Cao, & Wachowicz, 2021) accumulates event-triggered information from nine devices. The sensing machines associate with NodeMCU microcontroller employing Arduino. The Arduino script collects data of activated occasions. The details are multi-sensor documents, the kind of occasion, and the duration. Wi-Fi operating in 802.11 practices with the Wi-Fi frequency band of 2,4 GHz provides the capacity to transmit large quantities of knowledge. The algorithm depends on a post/subscribe procedure. The NodeMCU customer prints the event-triggered information to the MQTT dealer. It contributes to the subject. The indoor area was a workshop with distinctive characteristics like 32 processors, a projector, course platform, two entries, two warmth pumps, and numerous windows. The investigation assembles occurrences initiated by all the detectors attached to a device set in the center of the space and sampled for the highest scope.

1.3 Analytics model

Building an investigation standard (Runkler T. A., 2020) is a repetitive procedure of constructing new parameters, measuring them, and experimenting with them. Information examination is the methodology of scrutinizing and studying knowledge collection to conclude the details they keep. It explores the practices from unprocessed learning and obtaining helpful understandings from it. It allows enterprises to get real-time perspicuity about deals, commerce, economics, and derivative growth. It permits groups within corporations to cooperate and earn more consequences. It is beneficial for industries to research past interchange arrangements and balance prospective company procedures. It enables corporations to achieve a competitive benefit.

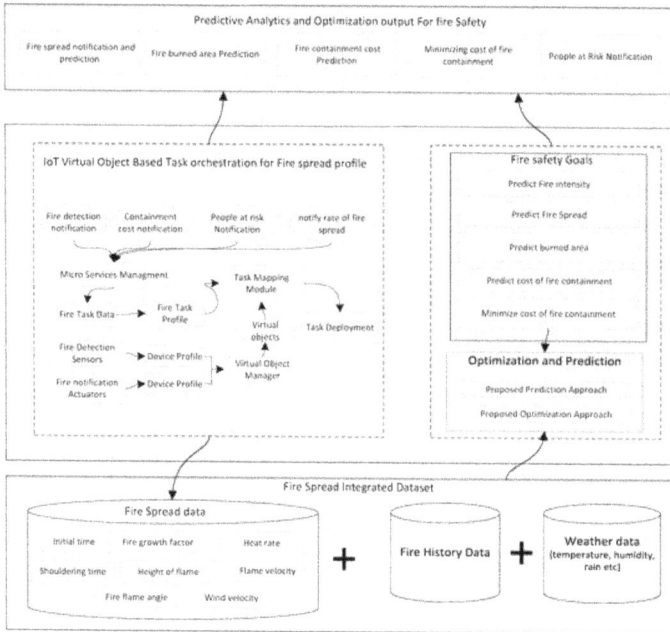

Figure 1.6 Design of proposed fire safety system (Iqbal, Ahmad, & Kim, 2021).

The suggested blaze security (Iqbal, Ahmad, & Kim, 2021) supplies benefits such as conflagration flared alarm, and its forecast. The initial section portrays the combined information collected from the blaze distributed details, its past, and climate facts. A conflagration planning supervisor utilizes duty plotting couples as an intake and delivers (balanced) methodical duos of these assignments and digital entities operating pace te-monotonic functions. Blaze protection schemes and simulated teams stow in the blaze duties' warehouse. The assignment distribution administrator makes operations from these composed couples and installs them on protection instruments. The machines' replies ship to the forecast phase for foretelling study and balancing. Outcomes from pre-prognostic ships to the shared management and protection division create containment plans. The IVToS developments are in the state of a conflagration stretch summary, including parameter details. The pre-trained collaborative representative indicates the blaze stretch summary. It also has a conflagration containment price based on the parameters. The safeness tier has objectives for unraveling issues of the security concerns slab. It calculates peak blaze spread. The physical tier organizes a blaze protection IoT warehouse. The conflagration spread tier possesses data from the blaze summary. The modeling phase delivers a mathematical preparation of blaze coverage and objective operation for underestimating the sum of blaze containment prices. IVToS application has registration details of the cliff citizens. The same is portrayed in the Figure 1.6.

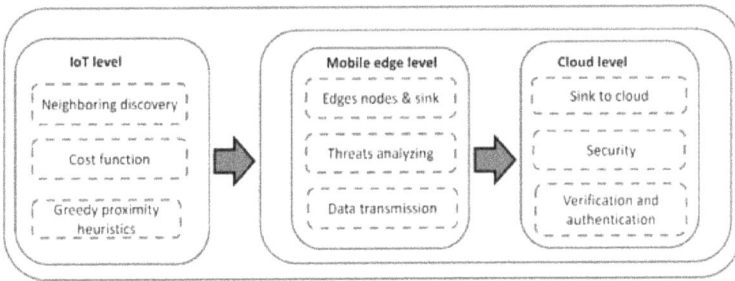

Figure 1.7 Components of the proposed medical model using IoT (Rehman, Haseeb, Saba, Lloret, & Tariq, 2021).

SBD-EC framework (Rehman, Haseeb, Saba, Lloret, & Tariq, 2021) has two stages. The therapeutic machines link in the form of a diagram. The separate vertex connects with corresponding neighbors by a price. The devices manage a regional controller that executes the operations of information accumulation. These middle machines seek to dispatch the healing details to the server and associate with stockpiles utilizing portable instruments. The SBD-EC model has three tiers. The detectors intercommunicate with the provincial arranger. The regional manager connects with the gateway employing different medium devices. The dynamic connectors cooperate with the warehouse. It lessens the latency in knowledge repository and computation. It delivers a protection procedure and transmission from transportable connectors to stockpile. It diminishes the transmission overhead and details latency. The framework employs greedy traversing with the closeness neighboring process to research trusted paths for shipping learning. Figure 1.7 depicts the same.

1.3.1 Classification model

A category example attempts to illustrate some findings from the intake values of exercise collection. It will foresee the type tags/varieties for the new knowledge.

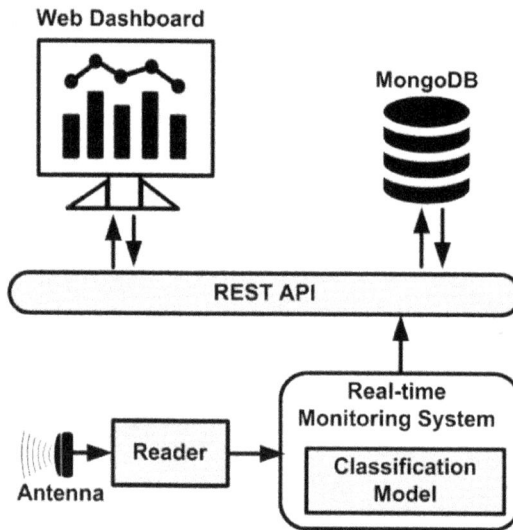

Figure 1.8 System architecture (Alfian, Syafrudin, Yoon, & Rhee, 2019).

The repository departure (Alfian, et al., 2020) has an RFID reading component and antenna before outcomes packing happens in the automobile. It precisely documents the titles using the entrance and stores them into the motor. The system jolts the favorable output. This unexpected reading happens when the labels discover within the read scope. The markers understand using the RFID reader utilizing the antenna. The system forwards the identification to the customer's system using the connected network. The knowledge congregation procedure accepts and proposes title data. It takes around 5 sec to drive the tramcar via the entrance. The architecture broadcasts related RSS facts on the screen and documented them in the assemblage sitting. The authors employed 86 × 54 × 1.8 mm inactive RFID labels for this test. It connects to the containers and the streetcar pushed via the entrance at various courses and paces. The frequency label has a frequency of 860–960 MHz. It uses Alien H3 and EPC Class1 Gen2 (ISO 18000-6C). The markers come with PVC fabric into the inactive title. The investigation employs a single reader ALR-9900+ from Alien Technology and linear antenna ALR-9610-AL with 5.90 dbi Gain. The working frequency of the reader was 902–928 MHz and backed EPC Class1 Gen2 (18000-6C). The ALR-9900+ is an industry reader, permitting customers to watch or read numerous markers. It accumulates RSS information at considerable lengths. The Alien reader supplies checklist. It is a full-characteristic approach for determining the identification of labels. The reader delivers various modulation methods and Gen2 protocol. The reader integrates different languages. Figure 1.8 depicts the same.

The suggestion (Tama & Rhee, 2017) manages a deep fuzzy system to categorize intrusion in the system. The protection architecture has three tiers. The perception tier owns the perception component employed to acquire facts. The layer addresses authorization issues between the devices. The transport slab has three tiers - the access grid, the core web, and the local location system. The tier delivers omnipresent admission data for the perception tier employing an unconnected system. A classifier is prepared to create a category representative from the attack facts. The work uses the same to make future predictions. The system uses UNSW-NB15 to validate the Intrusion detection system. It contains 42 characteristics. The architecture uses CIDDS-001 (tagged database) to authenticate the detection technique. Openstack with a gateway provides network traffic information. It has ten properties and five groups. The system uses 146500 instances to evaluate the work.

The suggested approach (Arridha, Sukaridhoto, Pramadihanto, & Funabiki, 2017) has three components. The learning procedure defines the construction of the category prototype. This methodology is done before the real-time category calibration. The precision class of the developed standard concerns the status of faith in category consequences. Real-time classification represents the design of the real-time investigation. The real-time category employs a detailed logical methodology. This procedure needs a high pace to be a stand-alone technology. It trims the hold between the information acceptance and envisioning. Real-time category operation employs Spark MLlib. The real-time illustration depicts the strategy on the front-end net client medium.

The centerpiece (Shirley, Sundari, Sheeba, & Rani, 2021) of the capacitive scanner is the condenser. The capacitance wiring manages data from fingerprints by forming a vision of the same. It has some primary components. The condenser is charge-keeping and links them to conductive plating. The visioner will allow us to maintain a trail of scanned particulars. A collection of capacitances is beneath the scanner to collect attributes of the object. The modifications regarded in the components follow with the service of Op-Amp combined wires. The frame stores the information using an A-D converter. The knowledge aggregated using the capacitive visioner is separated for individual entity scanned. The system compares the data with the storage data before initiating the motor. The computerized information explores to approximate its specific characteristics in the stockpile.

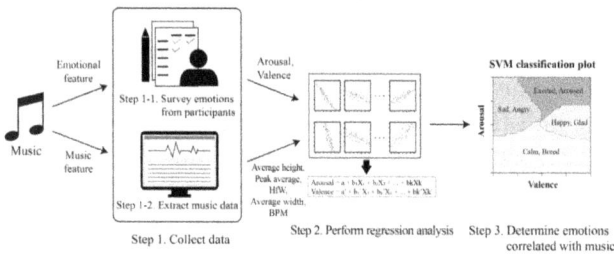

Figure 1.9 Overall procedure adopted (Seo & Huh, 2019).

The work (Seo & Huh, 2019) is feeling labeling research for melody. It mines the attribute information from the harmony. The study estimates passionate details connected with a tune. It organizes property-oriented facts. The authors explore the feelings linking with melody and analyze the sentiments based on two measurements-valence and arousal. They used the circumplex emotive framework. The features of the theme are under consideration. They define which feelings connect with the respective tune by studying the attributes. Regression examination concludes which feelings an individual senses when hearing to a specific theme. The authors employed 40 players in the study. Thirty-one parties were in their 20s, six were in their 30s, and three were in their 40s. The tune records accumulated 100 Korean pop melodies that earned international fame. The audio origin was an MP3 document with a grade of 320 kbps. The consequences of the sentiment category investigation of 35 songs were acquired by analyzing 40 players and assessed concerning the expressive variety emanated using the suggested methodology. Figure 1.9 represents the same.

The work (Reshma, Sathiyavathi, Sindhu, Selvakumar, & SaiRamesh, 2020) is a framework for ground conduct investigation and suggestions for crop growth. The pre-computation phase has two phases-information cleaning and missing details calibration. The design gives a discharge of each operation of an abstract prototype that describes the format and manners. The model is a computation connection with another in the atmosphere to forecast harvest. The details extraction method foretells the harvest for increasing yield efficiency. The pre-calibrated learning undergoes accumulation employing SVM and the Decision Tree procedure. The system predicts harvest yield based on the developed practices. The community and produce are intakes to the prophecy prototype. The system compares two processes, and precision is estimated to obtain the diagram. A methodological procedure measures different earth constraint. The system stores detailed information from the detectors into stockpiles, and data generates on a computerized output device. The method permits surveillance of ground states. The knowledge accumulates in real-time in the dimensions of 2000 x 1000 sq. feet. The area splits into three locations where one spot is 1000 x 1000 with good daylight and another is 500 x 500 openness to rays during dawn. The other location is in the dark. The observation of dampness, warmth, and water level is scrutinized constantly by linking the detectors with Raspberry Pi 3.

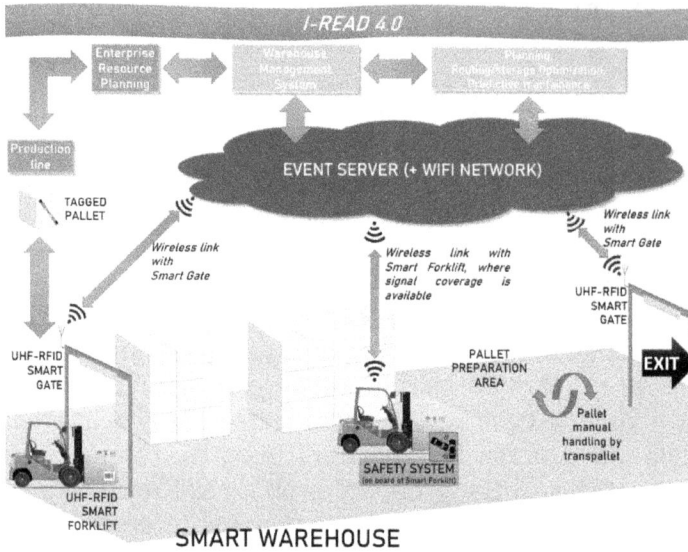

Figure 1.10 I-READ 4.0 framework (Motroni, Buffi, Nepa, Pesi, & Congi, 2021).

The suggestion (Motroni, Buffi, Nepa, Pesi, & Congi, 2021) utilizes the inactive UHF-RFID methodology. It is a combined grid of RFID readers, UHF-RFID intelligent admission, and UHF-RFID Smart Forklifts. It recognizes respective palettes, their situation, and their zone. The labeled entities are pallets having the outgrowth. The outcome is input to the repository maintained by forklifts. A forklift raises the pallet and sends it to the loading zone. The single pallet is packed onto the vehicle manually. The depository opening of the outcome and the doorway has UHF-RFID intelligent gateway. It watches the activities of derivatives and forklifts. The Forklift moves have a search strategy to permit real-time pallet localization. The pallet area connects to the forklift area at the time of the receiving occurrence. The server and Forklift transmit the information of the place and situation of a pallet to the storage hub. The facts of the position of pallets and forklifts authorize to build a real-time map of the repository. It allows to execution of a balancing procedure to enhance the administration. The details on the forklift location integrate with the details of the crash recognition method located on the forklift. It permits statistical investigation about the places with the danger of crashes. Figure 1.10 represents the same.

1.3.2 Clustering model

Clustering (Balakrishna, Solanki, Kumar, & Thirumaran, 2020) (Romesburg, 2004) (Kaufman & Rousseeuw, 2009) is an unconfirmed knowledge prototype. It batches information into collections based on the characteristics. It is one of the precomputing methodologies used earlier to categorize. The learning groups employ resemblance standards. It creates entity labels. It facilitates the role of the subject reviewer by delivering significant classes of facts. It uncovers the confidential design details and creates a set of groups of the identical variety in the database. It decreases the diagnosis price.

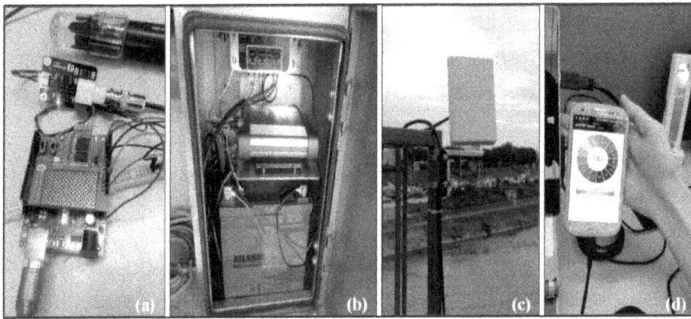

Figure 1.11 (a) Integrated Arduino-based Bluetooth shield pH sensor; (b) solar power-controlled battery (3 V to 5 V with 12 V to 24 V); (c) Bluetooth flat panel antenna (2.4 GHz with 15.5 dBi); and (d) Arduino-based mobile application for smartphones. (Jo & Baloch, 2017).

The representative location podium (Jo & Baloch, 2017) is in the Jungnangcheon rivulet of the Han tributary in Haengdang-dong, Seongdong-gu. The model estimates indistinct constraints at a stationary place for every interval. The study experimented for three months. The detector maps to Arduino shield V2.1. Some parameters measured include climate, DO, and pH. The model has stainless steel framework. The sensors used Solar fuel of 3 V to 5 V. The system transmits the energy of 12 V to 24 V to the Bluetooth antenna utilizing a seepage, decay, and water-proof storage repository. The model precomputes the acceptable limitations of water grade norms for river water in Korea in Arduino IDE 1.0.X. The work transmits the knowledge from the information accession unit to the hub. Bluetooth antenna aids in transmission. The architecture gets overview figures of periodic divergences in datasets. The study used a k-means clustering methodology. The system operated hierarchical grouping to organize matching couples and connect them in the next step. The work examined facts received from Jungnangcheon rivulet employing Density-based spatial clustering methodology. Figure 1.11 depicts the same.

The grid (Pita, Rodriguez, & Navarro, 2021)has acoustical devices installed in Barcelona. The system uses 86 acoustic detectors. The database delivers a long-term investigation. The machine senses the tension of its place in a constant method employing a Cesva TA120 remote sonometer. The precision follows International Standard IEC 61672-1. The device ships the package to the warehouse. It keeps and computes the information. The system computes long-term norms using acoustic stress status. The experiment uses k-means clustering, hierarchical agglomeration, partitioning around medoids, and expectation-maximization procedure.

1.3.3 Forecast model

The forecast approaches regulate measured value prognosis. It evaluates the values of recent information based on understandings from documented knowledge. It develops numerical texts in chrono-logical information (when data is unavailable).

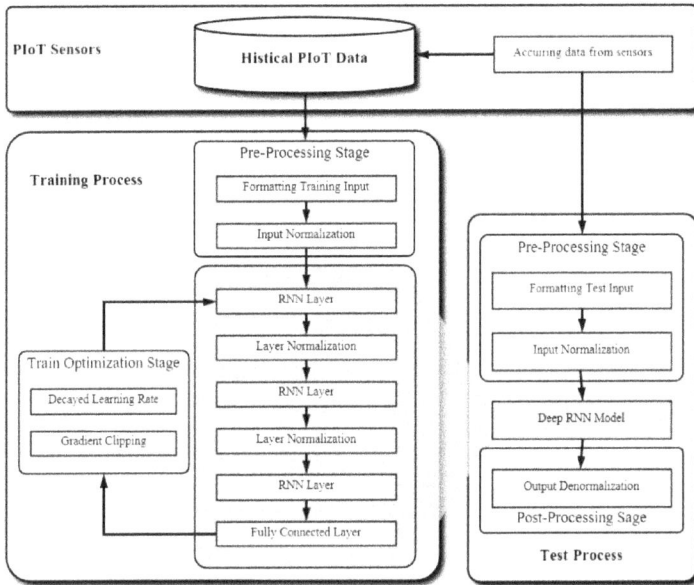

Figure 1.12 proposed model (Ahn & Park, 2021).

The PV energy podium (Ahn & Park, 2021) has ten serial-wired solar boards on the ceiling of the Engineering Construction in Konkuk University, Seoul, Korea. The architecture is a deep RNN with multiple slabs instances. The examination assesses the execution by employing the trial database. The investigation happens after finishing the workout stage. The system accepts intake from PIoT detectors. It has a PV fuel unit, solar radiation, ambient climate, moisture, and breeze pace. The system accumulates information occasionally using PIoT devices. The component forwards them to the pre-computation phase. The aggregated details structures to a demanded intermission to provide the infusion design. The formatted intake undergoes normalization. Figure 1.12 portrays the same. Table 1.1 summarizes the various contributions towards forecast model.

Table1.1 Contribution of various work towards forecast model

contribution	Characteristics	Hardware/software specification
IoT-based clever fish farming and trailing management (Gao, Xiao, & Chen, 2019)	• enables the collection, storage, analysis, prediction, tracking and querying of fishpond water-quality data • data forecasting methods based on the model tree algorithm • QR code label attached to aquatic product enables consumers to trace and examine historical farming process data	• Experiment was conducted in Asian carp and rainbow trout fish farm in the city of Baoding, Hebei Province, China. • area covered approximately 2 acres • The system has a temperature sensor (DS18B20), pH sensor (Shanghai Leici, Model E-201-C), dissolved oxygen sensor (YHT-8402), water electrical conductivity sensor (TDS) and water turbidity sensor (Water WT-RCOT) • GUI modules collected the fish farming environmental parameters
IoT-based system to forecast crop frost. (Guillén-Navarro, Pereñíguez-García, & Martínez-España, 2017)	• data acquisition network measures weather attributes directly taken in the crop field • public weather forecast service acts as a complementary source of information to implement our crop frost forecast service • Data Processing System is responsible for collecting weather data both from the crop field and the weather forecast service	• deployed to monitor a real crop of peach trees located in Cieza, a town located in south-east of Spain in Murcia region • hardware platform is an Arduino UNO • DS18B20 model measures temperature • DS3231 AT24C32 clock is used to obtain date and time
Drought Forecast and Alert System (Kung, Hua, & Chen, 2006)	• 4-tier system framework composed of Mobile Users (MUs), Ecology Monitoring Sensors (EMSs), Integrated Service Server (ISS), and Intelligent Drought Decision System (ID2 S) • monitor and collect continuously environmental drought data in combination with MODIS satellite images • monitors and collects all spatial and temporal ground surface information by using wireless sensor networks and network camera	• The system uses JAVA, Java 2 Platform. Micro Edition (J2ME), ArcGIS 8.0 and PostgreSQL • Hardware encompasses one MIB510, three MPR400, one MDA300 and one MTS420, Compaq Pads and network cameras
Traffic monitoring system (Xiao, Peng, Wang, Xu, & Hong, 2009)	• Gray forecast process original data and accumulate data into a Gray generated number. A differential equation and a response function for grey series forecast are established and forecast is carried out for Gray generated number and the last forecast value will be obtained after the final restore processing	• It is composed of ordinary nodes, sink nodes, gateway nodes and control centres • It uses Random Early Detection algorithm, Traffic Random Early Detection procedure
Solar energy forecast system (Dhillon, Madhu, Kaur, & Singh, 2020)	• weather information collected by sensors and three parameters from the input variables namely temperature, pressure and relative humidity are used for reference signal generation. • the features of the data are extracted using Independent Component Analysis algorithm and these features are fed to the feed forward neural network	• it uses ICA algorithm for feature extraction
UD-WCMA (Dehwah, Elmetennani, & Claudel, 2017)	• dynamical forecast scheme with adaptive parameters solves problem of pretuning arising in the standard solar power estimation techniques • It leverages a set of historical profiles, and fuses it with realtime data	• deployed 10 nodes over 6 months period to collect the measured data • motes are built around a 32- bit ARM Cortex M4 micro-controller with 192 kB RAM, running at a frequency of 168 MHz

1.3.4 Outlier model

It is a procedure where the investigation recognizes the unusual outcomes of an experiment. It removes erroneous observations in the collection.

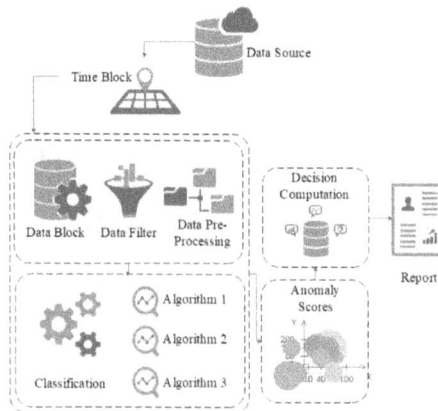

Figure 1.13 Detection Methodology (Hurst, Montañez, & Shone, 2020).

The sliding window procedure (Hurst, Montañez, & Shone, 2020) accumulates information into packages. It undergoes pre-computation to evaluate missing text. The outcome transmits using the outlier discovery process. The DBSCAN program recognizes groups based on the viscosity of the details in the characteristic pool. These promote precise labels and the split of the sets of various dimensions. The Local Outlier Factor strategy evaluates the strength of data and notices abnormal content by estimating the regional variation of information when approximated with its neighbors. OPTICS process perceives significant collections within datasets having differing intensities. Table 1.2 summarizes the contribution towards outlier analytics.

Table 1.2 contribution towards outlier model

contribution	Description	Experiment specification
Outlier mining (Mascıarı, 2007)	• The system detects anomalous data • the streams are related to a set of containers of tinned foods being tracked from the farm • The work uses encoding strategy • Fourier Transform is used compare different RFID streams	• 100 streams were evaluated. It was divided into 4 classes • Tuna fish readings had 26 tagged containers storing 500 cans • Tomato had readings giving 23 tagged containers storing 400 cans • Syrupy Peach readings gave 20 tagged containers storing 350 cans • Meat readings had 35 tagged containers storing 600 cans
(Basu & Meckesheimer, 2007)	• It is e two variations of a method that uses the median from a neighbourhood of a data point and a threshold value to compare the difference between the median and the observed data value. • first aspect is identifying which data points in a time series are outliers • It addresses the issue of what to do with a data point that has been identified as an outlier	• The work was simulated. • The outliers were generated for the 60 values
(Sheng, Li, Mao & Jiu, 2007)	• It is histogram- based detection approach to identify distance-based outliers	• The work is simulated using the datasets obtained from Intel lab. • The data were collected from 54 sensors during a one-month period
(Prastawa, Bullitt, Ho, & Gerig, 2004)	• It is a framework for automatic brain tumor segmentation from MR images • First stage - detect abnormal regions using a registered brain atlas as a model for healthy brains • Second stage - determine the intensity properties of the different tissue types • Third stage - apply geometric and spatial constraints to the detected tumor and edema regions	• The work used three real datasets. • The study used VALMET segmentation validation tool • 2 GHZ Intel Xeon machine is used.

1.3.5 Time series model

Time sequence investigation is a distinctive method of examining a series of information gathered throughout a period. The reviewers document facts at invariant gaps over a set period. It needs a large number of details to guarantee consistency and trustworthiness. It foretells based on chronological information. It enables institutions to comprehend the underlying reasons of systemic practices over the span.

Figure 1.14 Oktokopter in flight, fitted with Canon 550D camera (Turner, Lucieer, & De Jong, 2015).

The system (Turner, Lucieer, & De Jong, 2015) uses Oktokopter multi-rotor micro-UAV with a DroidWorx carbon fiber airframe, a MikroKopter robot pilot, and a Photoshop One camera gimbal. The Oktokopter has a load capability of 2 kg. figure 1.14 portrays the same. The aviation time is 5–10 min. It is alleviated camera support to preserve nadir images on the flight. MikroKopter aviation electrical techniques robotically sustain the grade of flying. It maintains the height log approach information at 1 Hz and mechanically flies the UAV using a sequence of predefined, 3D GPS points. They use Google earth picture. The system uses Canon 550D Digital Single Lens Reflex (DSLR) camera. The photo capture rate was maintained by the UAV's flying authority committee. It radiates a stimulus vibration at the selected frequency. The aviation authority panel is associated with a custom-made cord that activates the isolated shutter release of the camera. The camera works in a secure method. Pictures were caught in RAW structure and kept on the memory card in the camera. The aviation course was pre-computed into the UAVs mechanical pilot to flee a network practice over the avalanche.

1.4 Learning analytics

Learning Investigation is the measure, aggregation, examination, and documenting outcomes about pupils and their learning. The system interpretation and elevates education and the surroundings in which it materializes.

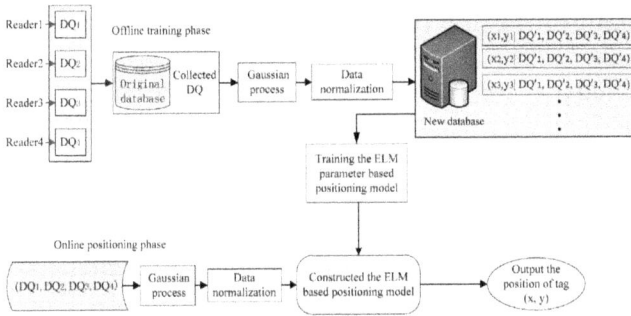

Figure 1.15 The framework of ELM IPS based on Gaussian process. (Wang, Shi, & Wu, 2017).

The system (Wang, Shi, & Wu, 2017) has two localization steps. The organized RSSI has relation labels and their physical surroundings having activity intakes and workout destinations (offline stage). The group category is guided for all DQ contents in the datasets using ELM knowledge. It determines the constraints of the grid. It demonstrates a primary ELM regression representative of background factors for online location. The real-time DQ text will be provided into the professional ELM standard in the online phase. Laird-S8658WPL UHF RFID network gathers the RFID sign. The experiment uses MATLAB with Windows 7 device with an Intel Core i7 processor of 3.07 GHz and 4 GB of RAM. Figure 1.15 portrays the same. Table 1.3 summarizes the contributions towards learning data analytics.

Table 1.3 Contribution towards Learning data analytics

Contribution	Description of work	Experiment specification
(Sliwa, Piatkowski, & Wietfeld, 2020)	• LIMITS framework supports data analysis and code generation • LIMITS is designed as a novel open-source machine learning framework for IoT applications, which provides automation features for high-level data analysis tasks and platform-specific code generation • The work integrated automatic static Worst-case Execution Time (WCET) analysis into the model selection process	• cellular data rate prediction in vehicular networks and radio-based vehicle classification were case studies considered in the work
(Hitimana, Bajpai, Musabe, Sibomana, & Kayalvizhi, 2021)	• A global positioning system (GPS) is used to locate the building • temperature sensor, humidity sensor, lighting sensor, CO2 sensor, and the passive infrared sensor were used in the trails • At the fog node, all possible collected data is stored in the datacentre of the institution for real-time analysis/data processing by reducing the latency of the wide-area network • At the cloud node, third-party cloud services are selected, allowing public data access for decision-making followed by prediction analysis	• implemented in the University of Rwanda building located in Kigali city, Rwanda • sensor modules are connected to an ESP 8266 microcontroller NodeMCU
(Singh, Cha, Kim, & Park, 2021)	• The work is ELM based big data analysis in the cloud layer for IoT • The study uses PCA algorithm for feature extraction, K-means algorithm for scaling, and Naïve Bayes algorithm for classification in the edge layer	• NSL-KDD - KDDTest+, KDDTest-21, and KDDTrain+ dataset was adopted • Weka tool is used to estimate the experimental analysis • The work used Processor E5-1620 v3
(Shakeel, et al. 2021)	• The work targets the detection of root servers causing vulnerabilities • It implements blockchain technology for access log and control management • adversary event detected by logistic regression is filtered using cross-validation to retain the precision of data analysis for varying user density and virtualized resources	• The work is evaluated using Contiki Kooja simulator • 120 IoT devices accessing 48 VRs were used • The system had 10 resource servers of size 1TB each
(Onal, Sezer, Ozbayoglu, & Dogdu, 2017)	• The algorithm has Data acquisition phase, Extraction, Transform, Load and semantic phase, and learning phase.	• LinkedSensorData and LinkedObservationData were used - 8000 weather sensors
(Babu, et al. 2016)	• During the check-in phase, each bag is attached with a passive RFID tag which is to be detected by the deployed RFID readers • the bags enter the main sorting area (SO) where the sorting system ensures that the bag is pushed into a designated chute • The bags are then loaded into wagons by the baggage handling staff, before they are transported to a designated aircraft through one of the gateways • bags pass through a belt-loader reader while loading them into a plane	• real data set collected from an airport that operates an automatic RFID-based baggage handling system • the system uses 20,000 RFID tagged bags

1.4.1 Collaborative and interactive learning

The Knowledge investigation Collective is a corporation among en-lightening investigators, information researchers, and visionary leaders from universities. It is knowledge covering ecosphere look-ing for to connect the authority of Education Analytics to examine and distribute learning of instruction and scholarship in the con-ducts that lead to improved inquiries, deeper thinking, and more data delivery and conclusions for fruitful schooling.

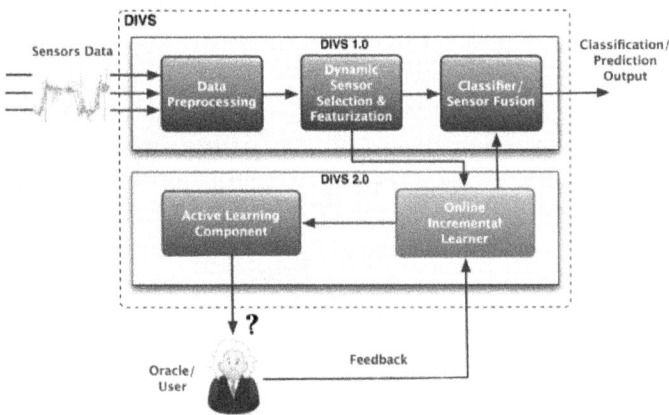

Figure 1.16 DIVS data processing pipeline (Tegen, Davidsson, Mihailescu, & Persson, 2019).

The work (Tegen, Davidsson, Mihailescu, & Persson, 2019) presents a progressive interpretation of the DIVS vision by incorporating a collaborative apparatus education. It allows the design to take in-take from customers and the environment. The architecture is a va-riety detector database experiencing a knowledge combination and category approach. It trails active measurement known as input re-turns utilizing learning delivered by sensing devices. The unit evalu-ates the effectiveness of the detector by tallying the expected out-come. The system enhances the effectiveness and scalability of the current prototype acquainted in the conventional collection in un-derstanding manner with a labeled knowledge collection. It ad-dresses problems where knowledge reaches over the period and has no tagged information is unrestricted. The investigations employ sequences from devices to assess the tenancy of an enclosure. The recordset includes a registered series of learning representatives from five characteristics. The measurement of the information dataset is between 2664 illustrations and 9752 samples. The authors utilize the weighted Naïve Bayes procedure to categorize data and the Variable Uncertainty Method for engaged education. Figure 1.16 portrays the same. Table 1.4 summarizes the contribution towards collaborative and interactive learning.

Table 1.4 contribution of various authors towards collaborative and interactive learning

Contribution	Description of work	Experiment specifications
(Tan, Liu, & Chang, 2007)	• The author offers an Environment of Ubiquitous Learning with Educational Resources (EULER) based on RFID, the Internet, ubiquitous computing, embedded devices, and database technologies. • Teachers and kids from elementary schools were among the participants. • The EULER has two phases. The mobile-Based Interactive Learning Environment server for teachers and mobile tools for students. Teachers must first add teaching materials to the Mobile Content Database and exams before teaching. During class, the MOBILE server handles requests from students and responds to them. • The unit creates a virtual classroom and enables a variety of learning activities such as bulletin boards, forums, voting, chatting, assignments, assessment, and instruction. • After the outdoor teaching is over, teachers can use a Mobile Assessment Management unit to administer assessments and analyse student performance. • The Mobile Learning Record Database saves student learning statuses and records, such as assignment grades, reading times, the number of discussions in which they have participated, instances of data gathering, and instances of information sharing.	• 5 grades are considered. • Teachers have at least 10 years of experience. • The study used Cronbach coefficient to portray efficiency.
(Huang, Chang, & Sandnes, 2010)	• The paper explains the architecture of the proposed guide system, which includes a PDA-based recommendation guide for art museums and a Radiofrequency Identification-based interactive learning system for science and engineering education that uses collaborative filtering technology. • The interactive kiosks and the central server make up the Hands-Free Interaction Guide System. • The learning service provides training materials and examinations based on the visitor's profile, while the network verification layer assures a reliable and durable connection.	• The work used 8000 sessions. • The study covered 20 different science subjects are distributed across three floors. • It had 20 kiosks are connected to single server.
(Eberle, et al. 2013)	• RFID trackers were used to track participants' interaction behaviour, and social network questionnaires and a bibliographic analysis offered further information. • It's a quasi-experimental design, the aspect of community knowledge support differed in workshops, and the factor community engagement level varied spontaneously among participants. The gadgets trace face-to-face closeness. • The system issued questionnaires to participants to fill out to express their thoughts. • The work examined the bibliography.	• software used HLM 6.08 • 152 persons participated • 8 workshops and a doctoral consortium
(Egusa, et al. 2016)	• members experienced both elements of the cooperative intuitive learning game. • The conducted poll comprised of a sum of six things. • Three things connected with the experience of cooperative communication with the leap work and the excess three things zeroed in on the opportunity for growth of the game.	• 18 primary school children with deafness, from third to sixth grade
(Sam, Roy, Akhter, & Whaiduzzaman, 2018)	• The work is a web of things based mixed learning • It is an approach that can upgrade customary instruction framework with (imaginative) learning techniques and advancements. • The review is a level-based methodology. • instructor plan course layouts of this course. • IoT brilliant gadgets give substance to understudies through a substance conveyance organization. • Brilliant learning incorporates the conveyance of notes and class substance. • Brilliant Pedagogy is a bunch of showing procedures, exercises, and decisions to comprehend the student objectives. • RFID empowered advanced library get important books as our prerequisites. • Customized learning is criticism-based instructing and learning climate customizes to individual understudies.	

1.4.2 Behavior learning

It is the interdisciplinary domain. It integrates approaches from the behavioral sciences, such as psychology, economics, sociology, and business. It incorporates computational techniques, statistics, information-centric engineering, knowledge procedures analysis, and mathematics. It aims to provide a better benchmark, comprehend and anticipate conduct.

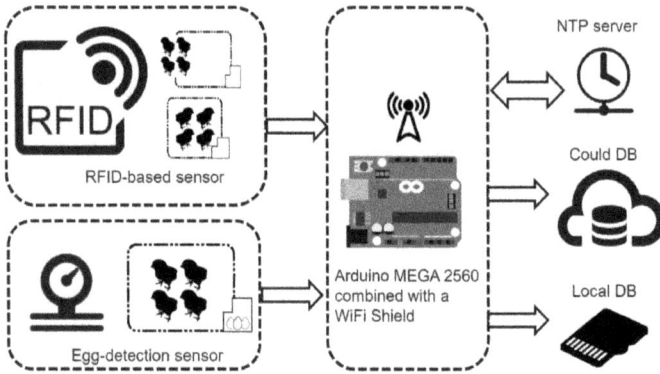

Figure1.17 IoT platform (Chien & Chen, 2018).

The work (Chien & Chen, 2018) carried out an RFID-based IoT system involving Arduino as a microcontroller unit and WiFi safeguard to empower admittance to a cloud information base and the organization time convention (NTP) server. The animals wore an RFID tag on one leg, and an RFID peruse was introduced in each home box to empower the discovery of individual hens entering or leaving the home box. The system coordinated the heap cells. The home box checks the laying of eggs by hens continuously. Figure 1.17 portrays the same. Table 1.5 represents the contribution towards behavior learning.

Table1.5 contribution towards behavior learning

Contribution	Description of work	Experiment specifications
(Tao, Shaik, Higgins, Gururajan, & Zhou, 2021)	• The review re-enacts an emergency clinic for constant human information assortment and constructs AI models to investigate the information designs. • The structure with two levels, isolating the information assortment and information demonstrating perspectives. • Two UHF 870 RFID peruse receiving wires were in chosen positions in the re-enacted ward that got information from the static label. • The recorded directions of the peruse receiving wire positions and title ascertain the distances between the tag. The calculation utilized conditions. • AI calculations were then consequently picked for demonstrating and information examination, planning to distinguish the relationship and foresee name values. • The system used a Group Learning strategy	• 16 arbitrarily chosen reader-antenna positions resulted in 120 combinations of the paired readers. • mean absolute error and mean squared error was computed.
(Mansbridge, et al, 2018)	• The work used an aggregate of six sheep from 140 creatures at the University of Nottingham. • The gadgets connect to the six sheep in two areas. • The system gathered attributes information. Body condition scoring of lamb evaluates the level of largeness and body state of the living creature. • The sheep were kept in a rectangular, 0.3-section of the land field with a 179.3 m edge during the daytime. • It was permitted into a bigger 2.1-section of the land field (until next morning) while recording began again around evening.	• Sheep behavioural activities were recorded using a handheld Panasonic HC-V380 video camera with a tripod and were time stamped. • The video camera was fitted with a 64 GB SanDisk elite SDXC UHS-1SD card to store the footage. • Python 3.5 was used • random forest, support vector machine, k-nearest neighbour and adaptive boost was adopted.

1.4.3 Personalized learning

Personalized education benefits schooling, professional colleges, specialized workout universities. It reaches better heights by altering the scholar's understanding of the goal to fit their distinctive knowledge requirements. It enhances their education venture by giving replies to queries in a learned structure, teaching them through an investigation of a case, and suggesting or revisiting some topics of importance.

Figure 1.18 AttentivU System (Kosmyna & Maes, 2019).

The system (Kosmyna & Maes, 2019) uses vibrotactile response. A shawl works as a form-aspect for the comment module. The architecture estimates the rendezvous catalogue signifying the changes in the action of the brain. The output directories use multiple situation and assignments. It trails, watches, repositories administration and transmission chore. The skull strip assembles information using Wi-Fi. Figure 1.18 portrays the same.

1.4.4 Social learning and network-based analytics

Sociable grid investigation is known as web science. It is a domain of information examination that utilizes grids and diagram approaches to comprehend colonial systems. It is suitable for connections beyond the societal kingdom.

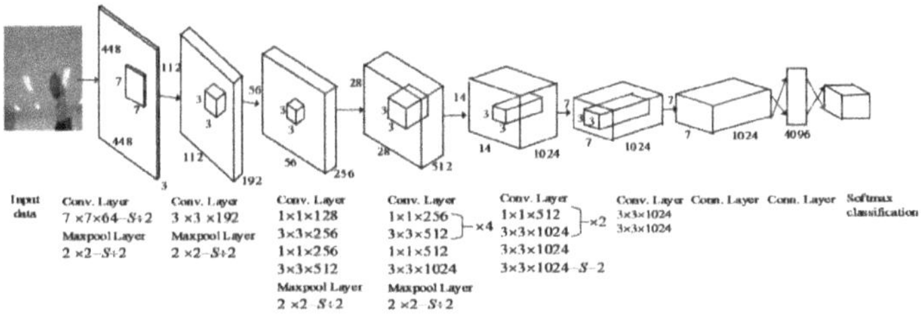

Figure1.19 Neural network coordinates (Yang, Yang, Sheng, Junior, & Li, 2018).

YOLO (Yang, Yang, Sheng, Junior, & Li, 2018) is a GoogLeNet frame-work-inspired real-time entity recognition prototype. It delivers an understanding of the intake picture and the things in the image. It can deliver end-to-end exercises and witnesses entities in real-time with moderate accuracy. The beginning convolutional tiers of the grid reaction mine characteristics from the infusion picture, and its completely interconnected ones complete the outcome possibilities and align. This structure has 24 convolutional slabs and two linked slabs. It consecutively employs smaller convolutional slabs to de-crease the attribute distance from the preceding tiers. The compartment answers to notice the entity (when the system discovers the middle of the entity in a particular network division). It utilizes a five-variant to describe the boundaries. The trust-weighs validates the standard of partition comprising entity and how accurate it de-fines the boundary it indicates. Figure 1.19 represents the same. Table 1.6 summarizes the contribution towards social learning and network-based analytics.

Table 1.6 contribution towards social learning and network-based analytics

Contribution	Description of work	Experiment specifications
(Benitez-Andrades, Arias, Garcia-Ordás, Martinez-Martinez, & Garcia-Rodríguez, 2020)	• The research demonstrates the viability of an eHealth solution for addressing teenage food patterns and physical activity. • Healthy eating habits and physical activity levels were investigated. • The body measurements of the students served as the study's variables.	• The sample consisted of 139 adolescents • SPSS v 24.0 software was used
(Qiu, Qiu, & Lu, 2020)	• The solution employs traditional selective encryption techniques by encrypting only a portion of the input data. • A wavelet technique compresses ECG data recorded at the sensors' endpoints. • Based on supervised machine learning-based classification, the experiment rethinks the reference levels of the wavelet frequency bands.	• Three separate cardiac disorders, each with its own set of symptoms, are represented by ECG signals. • Support Vector Machine (SVM) is employed.
(Li et al., 2018)	• To achieve RFID indoor localization, an Artificial Immune System-based Radial Basis Function Neural Network (AIS-RBF-NN) is proposed. • An artificial immune system is employed to optimize radial function center vector selection. • There are two steps to it. The data groups supply to the AIS-RBF-NN model in the offline stage. By assessing the AIS-RBF-NN model in the online stage, the real-time RSSI values and their differences from the target tag are fed into the trained AIS-RBF-NN, and the resulting output is the estimated coordinates of the target tag.	• The experiment uses 10m *10m area.

References

Abiteboul, S. (1997). Querying semi-structured data. *International Conference on Database Theory* (pp. 1-18). Greece: Springer, Berlin, Heidelberg.

Ahn, H. K., & Park, N. (2021). Deep RNN-based photovoltaic power short-term forecast using power IoT sensors. *Energies, 14*(2), 436.

Alfian, G., Syafrudin, M., Farooq, U., Ma'arif, M., Syaekhoni, M., Fitri-yani, N., . . . Rhee, J. (2020). Improving efficiency of RFID-based traceability system for perishable food by utilizing IoT sensors and machine learning model. *Food Control, 110,* 107016.

Alfian, G., Syafrudin, M., Yoon, B., & Rhee, J. (2019). False positive RFID detection using classification models. *Applied Sciences, 9*(6), 1154.

Alqahtani, M., Gumaei, A., Mathkour, H., & Maher Ben Ismail, M. A. (2019). A Genetic-Based Extreme Gradient Boosting Model for Detecting Intrusions in Wireless Sensor Networks. *Sensors, 19,* 4383.

Arridha, R., Sukaridhoto, S., Pramadihanto, D., & Funabiki, N. (2017). Classification extension based on IoT-big data analytic for smart environment monitoring and analytic in real-time system. *International Journal of Space-Based and Situated Computing, 7*(2), 82-93.

Arumugam, P., Chemura, A., Schauberger, B., & Gornott, C. (2021). Remote Sensing Based Yield Estimation of Rice (Oryza Sativa L.) Using Gradient Boosted Regression in India. *Remote Sens, 13,* 2379.

Awan, A. J., Brorsson, M., Vlassov, V., & Ayguade, E. (2015). *How data volume affects spark based data analytics on a scale-up server.* Cham: Springer.

Baba, A. I., Jaeger, M., Lu, H., Pedersen, T. B., Ku, W. S., & Xie, X. (2016). Learning-based cleansing for indoor rfid data. *International Conference on Management of Data* (pp. 925-936). San Francisco California USA: ACM.

Balakrishna, S., Solanki, V. K., Kumar, R., & Thirumaran, M. (2020). Survey on machine learning-based clustering algorithms for iot data cluster analysis. In S. V, L. Z. Hoang M., & P. P. (eds), *Intelligent Computing in Engineering* (pp. 1195-1204). Singapore.: Springer.

Basu, S., & Meckesheimer, M. (. (2007). Automatic outlier detection for time series: an application to sensor data. *Knowledge and Information Systems, 11*(2), 137-154.

Benítez-Andrades, J. A., Arias, N., García-Ordás, M. T., Martínez-Martínez, M., & García-Rodríguez, I. (2020). Feasibility of social-network-based eHealth intervention on the improvement of healthy habits among children. *Sensors, 20*(5), 1404.

Berkhin, P. (2006). A survey of clustering data mining techniques. In K. J., N. C., & T. M. (eds), *Grouping multidimensional data* (pp. 25-71). Berlin, Heidelberg: Springer.

Chien, Y. R., & Chen, Y. X. (2018). An RFID-based smart nest box: an experimental study of laying performance and behavior of individual hens. *Sensors, 18*(3), 859.

Dehwah, A. H., Elmetennani, S., & Claudel, C. (2017). UD-WCMA: An energy estimation and forecast scheme for solar powered wireless sensor networks. *Journal of Network and Computer Applications, 90*, 17-25.

Dhillon, S., Madhu, C., Kaur, D., & Singh, S. (2020). A solar energy forecast model using neural networks: Application for prediction of power for wireless sensor networks in precision agriculture. *Wireless Personal Communications, 112*(4), 2741-2760.

Eberle, J., Stegmann, K., Lund, K., Barrat, A., Sailer, M., & Fischer, F. (2013). Fostering Learning and Collaboration in a Scientific Community-Evidence from an Experiment Using RFID Devices to Measure Collaborative Processes. *CSCL 2013* (pp. 169-175). Madison, WI, USA: International Society of the Learning Sciences.

Egusa, R., Sakai, T., Tamaki, H., Kusunoki, F., Namatame, M., Mizoguchi, H., & Inagaki, S. (2016). Preparatory development of a collaborative/interactive learning game using bodily movements for deaf children. *15th International Conference on Interaction Design and Children* (pp. 649-653). Manchester United Kingdom: ACM.

Eletter, S., Yasmin, T., Elrefae, G., Aliter, H., & & Elrefae, A. (2020). Building an Intelligent Telemonitoring System for Heart Failure: The Use of the Internet of Things, Big Data, and Machine Learning. *21st International Arab Conference on Information Technology (ACIT)* (pp. 1-5). Giza, Egypt: IEEE.

Elsaleh, T., Enshaeifar, S., Rezvani, R., Acton, S. T., & Janeiko, V. &.-E. (2020). IoT-Stream: A lightweight ontology for internet of things data streams and its use with data analytics and event detection services. *Sensors, 20*(4), 953.

Gao, G., Xiao, K., & Chen, M. (2019). An intelligent IoT-based control and traceability system to forecast and maintain water quality in freshwater fish farms. *Computers and Electronics in Agriculture, 166*, 105013.

Goebel, M., & Gruenwald, L. (1999). A survey of data mining and knowledge discovery software tools. *ACM SIGKDD explorations newsletter, 1*(1), 20-33.

Guillén-Navarro, M. Á., Pereñíguez-García, F., & Martínez-España, R. (2017). IoT-based system to forecast crop frost. *International Conference on Intelligent Environments (IE)* (pp. 28-35). Seoul, Korea: IEEE.

Hitimana, E., Bajpai, G., Musabe, R., Sibomana, L., & Kayalvizhi, J. (2021). Implementation of IoT framework with data analysis using deep learning methods for occupancy prediction in a building. *Future Internet, 13*(3), 67.

Huang, Y. P., Chang, Y. T., & Sandnes, F. E. (2010). Experiences with RFID-based interactive learning in museums. *International Journal of Autonomous and Adaptive Communications Systems, 3*(1), 59-74.

Hurst, W., Montañez, C. A., & Shone, N. (2020). Time-pattern profiling from smart meter data to detect outliers in energy consumption. *IoT, 1*(1), 92-108.

Iqbal, N., Ahmad, S., & Kim, D. H. (2021). Towards mountain fire safety using fire spread predictive analytics and mountain fire containment in iot environment. *Sustainability, 13*(5), 2461.

Javeed, M., Jalal, A., & Kim, K. (2021). Wearable Sensors based Exertion Recognition using Statistical Features and Random Forest for Physical Healthcare Monitoring. *International Bhurban Conference on Applied Sciences and Technologies (IBCAST)* (pp. 512-517). Islamabad, Pakistan: IEEE.

Jo, B., & Baloch, Z. (2017). Internet of things-based arduino intelligent monitoring and cluster analysis of seasonal variation in physicochemical parameters of Jungnangcheon, an urban stream. *Water, 9*(3), 220.

Kambatla, K., Kollias, G., Kumar, V., & Grama, A. (2014). Trends in big data analytics. *Journal of parallel and distributed computing, 74*(7), 2561-2573.

Kaufman, L., & Rousseeuw, P. J. (2009). *Finding groups in data: an introduction to cluster analysis.* Hoboken, New Jersey: John Wiley & Sons.

Kaur, P., Kumar, R., & Kumar, M. (2019). A healthcare monitoring system using random forest and internet of things. *Multimedia Tools and Applications, 78*(14), 19905-19916.

Khanum, M., Mahboob, T., Imtiaz, W., Ghafoor, H. A., & Sehar, R. (2015). A survey on unsupervised machine learning algorithms for automation, classification and maintenance. *International Journal of Computer Applications, 119*(13), 34-39.

Kim, C., Lee, H., Devaraj, V., Kim, W., Lee, Y., Kim, Y., . . . Sun, H. (2020). Hierarchical cluster analysis of medical chemicals detected by a bacteriophage-based colorimetric sensor array. *Nanomaterials, 10*(1), 121.

Kosmyna, N., & Maes, P. (2019). Attentivu: An EEG-based closed-loop biofeedback system for real-time monitoring and improvement of engagement for personalized learning. *Sensors, 19*(23), 5200.

Kumar, K. S., & Chezian, R. M. (2012). A survey on association rule mining using apriori algorithm. *International Journal of Computer Applications, 45*(5), 47-50.

Kung, H. Y., Hua, J. S., & Chen, C. T. (2006). Drought forecast model and framework using wireless sensor networks. *Journal of information science and engineering, 22*(4), 751-769.

Lai, X., Yang, T., Wang, Z., & Chen, P. (2019). IoT Implementation of Kalman Filter to Improve Accuracy of Air Quality Monitoring and Prediction. *Appl. Sci., 9*, 1831.

Li, Z., He, G., Li, M., L., M., Chen, Q., H. J., . . . Wang, S. (2018). RBF neural network based RFID indoor localization method using artificial immune system. *Chinese Control And Decision Conference (CCDC)* (pp. 2837-2842). Shenyang, China: IEEE.

Liu, X., & Pan, R. (2021). Boost-R: Gradient boosted trees for recurrence data. *Journal of Quality Technology, 53*(5), 545-565.

Mansbridge, N., Mitsch, J., Bollard, N., Ellis, K., Miguel-Pacheco, G. G., & Dottorini, T. &. (2018). Feature selection and comparison of machine learning algorithms in classification of grazing and rumination behaviour in sheep. *Sensors, 18*(10), 3532.

Masciari, E. (2007). A Framework for Outlier Mining in RFID data. *11th International Database Engineering and Applications Symposium (IDEAS 2007)* (pp. 263-267). Banff, AB, Canada: IEEE.

Mohnen, S. M., Rotteveel, A. H., Doornbos, G., & Polder, J. J. (2020). Healthcare expenditure prediction with neighbourhood variables–a random forest model. *Statistics, Politics and Policy, 11*(2), 111-138.

Motroni, A., Buffi, A., Nepa, P., Pesi, M., & Congi, A. (2021). An Action Classification Method for Forklift Monitoring in Industry 4.0 Scenarios. *Sensors, 21*(15), 5183.

Muhammad, I., & Yan, Z. (2015). SUPERVISED MACHINE LEARNING APPROACHES: A SURVEY. *ICTACT Journal on Soft Computing, 5*(3), 946-952.

Nelder, J. A., & Wedderburn, R. W. (1972). Generalized linear models. *Journal of the Royal Statistical Society: Series A (General), 135*(3), 370-384.

North, K. (1997). *Environmental business management: an introduction.* (Vol. 30). Geneva, Switzerland: International Labour Organization.

Onal, A. C., Sezer, O. B., Ozbayoglu, M., & Dogdu, E. (2017). Weather data analysis and sensor fault detection using an extended IoT framework with semantics, big data, and machine learning. *IEEE International Conference on Big Data (Big Data)* (pp. 2037-2046). Boston, MA, USA: IEEE.

Parise, A., Manso-Callejo, M. A., Cao, H., & Wachowicz, M. (2021). Prophet model for forecasting occupancy presence in indoor spaces using non-intrusive sensors. *AGILE: GIScience Series, 2,* 1-13.

Park, S. H., Lee, D. G., Park, J. S., & Kim, J. W. (2021). A Survey of Research on Data Analytics-Based Legal Tech. *Sustainability, 13*(14), 8085.

Pita, A., Rodriguez, F. J., & Navarro, J. M. (2021). Cluster Analysis of Urban Acoustic Environments on Barcelona Sensor Network Data. *International Journal of Environmental Research and Public Health, 18*(16), 8271.

Prastawa, M., Bullitt, E., Ho, S., & Gerig, G. (2004). A brain tumor segmentation framework based on outlier detection. *Medical image analysis, 8*(3), 275-283.

Qiu, H., Qiu, M., & Lu, Z. (2020). Selective encryption on ECG data in body sensor network based on supervised machine learning. *Information Fusion, 55,* 59-67.

Rehman, A., Haseeb, K., Saba, T., Lloret, J., & Tariq, U. (2021). Secured Big Data Analytics for Decision-Oriented Medical System Using Internet of Things. *Electronics, 10,* 1273.

Reshma, R., Sathiyavathi, V., Sindhu, T., Selvakumar, K., & Sai-Ramesh, L. (2020). IoT based classification techniques for soil content analysis and crop yield prediction. *Fourth International Conference on I-SMAC (IoT in Social, Mobile, Analytics and Cloud)(I-SMAC)* (pp. 156-160). Palladam, India: IEEE.

Romesburg, C. (2004). *Cluster analysis for researchers.* Morrisville, North Carolina, United States: Lulu.com.

Runkler, T. A. (2016). *Data analytics.* Fachmedien Wiesbaden: Springer.

Runkler, T. A. (2020). *Data analytics.* Fachmedien Wiesbaden.: Springer.

Satu, M. S., Roy, S., Akhter, F., & Whaiduzzaman, M. (2018). IoLT: an IOT based collaborative blended learning platform in higher education. *International Conference on Innovation in Engineering and technology (ICIET)* (pp. 1-6). Dhaka, Bangladesh: IEEE.

Seo, Y. S., & Huh, J. H. (2019). Automatic emotion-based music classification for supporting intelligent IoT applications. *Electronics, 8*(2), 164.

Shaik, A. B., & Srinivasan, S. (2019). A brief survey on random forest ensembles in classification model. *International Conference on Innovative Computing and Communications* (pp. 253-260). New Delhi, India: Springer, Singapore.

Shakeel, P. M., Baskar, S., Fouad, H., Manogaran, G., Saravanan, V., & Montenegro-Marin, C. E. (2021). Internet of things forensic data analysis using machine learning to identify roots of data scavenging. *Future Generation Computer Systems, 115*, 756-768.

Sheng, B., Li, Q., Mao, W., & Jin, W. (2007). Outlier detection in sensor networks. *8th ACM international symposium on Mobile ad hoc networking and computing* (pp. 219-228). Montreal Quebec Canada: ACM.

Shirley, D., Sundari, V. K., Sheeba, T. B., & Rani, S. S. (2021). Analysis of IoT-enabled intelligent detection and prevention system for drunken and juvenile drive classification. In K. M., & N. R. (eds), *Automotive Embedded Systems* (pp. 183-200). Cham: Springer.

Singh, S. K., Cha, J., Kim, T. W., & Park, J. H. (2021). Machine learning based distributed big data analysis framework for next generation web in IoT. *Computer Science and Information Systems, 18*(2), 597-618.

Sliwa, B., Piatkowski, N., & Wietfeld, C. (2020). LIMITS: Lightweight machine learning for IoT systems with resource limitations. . *IEEE International Conference on Communications (ICC)* (pp. 1-7). Dublin, Ireland: IEEE.

Sun, S., Bi, J., Guillen, M., & Pérez-Marín, A. M. (2020). Assessing driving risk using internet of vehicles data: An analysis based on generalized linear models. *Sensors, 20*(9), 2712.

Tama, B. A., & Rhee, K. H. (2017). Attack classification analysis of IoT network via deep learning approach. *Briefs Inf. Commun. Technol. Evol. (ReBICTE), 3,* 1-9.

Tan, T. H., Liu, T. Y., & Chang, C. C. (2007). Development and evaluation of an RFID-based ubiquitous learning environment for outdoor learning. *Interactive Learning Environments, 15*(3), 253-269.

Tan, X., Su, S., Huang, Z., Guo, X., Zuo, Z., Sun, X., & Li, L. (2019). Wireless sensor networks intrusion detection based on SMOTE and the random forest algorithm. *Sensors, 19*(1), 203.

Tao, X., Shaik, T. B., Higgins, N., Gururajan, R., & Zhou, X. (2021). Remote patient monitoring using radio frequency identification (RFID) technology and machine learning for early detection of suicidal behaviour in mental health facilities. *Sensors, 21*(3), 776.

Tegen, A., Davidsson, P., Mihailescu, R. C., & Persson, J. A. (2019). Collaborative sensing with interactive learning using dynamic intelligent virtual sensors. *Sensors, 19*(3), 477.

Tsai, C. W., Lai, C. F., Chao, H. C., & Vasilakos, A. V. (2015). Big data analytics: a survey. *Journal of Big data, 2*(1), 1-32.

Turner, D., Lucieer, A., & De Jong, S. M. (2015). Time series analysis of landslide dynamics using an unmanned aerial vehicle (UAV). *Remote Sensing, 7*(2), 1736-1757.

Vilalta, R., Apte, C. V., Hellerstein, J. L., Ma, S., & Weiss, S. M. (2002). Predictive algorithms in the management of computer systems. *IBM Systems Journal, 41*(3), 461-474.

Wang, C., Shi, Z., & Wu, F. (2017). Intelligent RFID indoor localization system using a Gaussian filtering based extreme learning machine. *Symmetry, 9*(3), 30.

Xiao, L., Peng, X., Wang, Z., Xu, B., & Hong, P. (2009). Research on traffic monitoring network and its traffic flow forecast and congestion control model based on wireless sensor networks. *International Conference on Measuring Technology and Mechatronics Automation. 1,* pp. 142-147. Zhangjiajie, China: IEEE.

Yafooz, W. M., Abidin, S. Z., Omar, N., & Idrus, Z. (2013). Managing unstructured data in relational databases. *IEEE Conference on Systems, Process & Control (ICSPC)* (pp. 198-203). Kuala Lumpur, Malaysia: IEEE.

Yang, G., Yang, J., Sheng, W., Junior, F. E., & Li, S. (2018). Convolutional neural network-based embarrassing situation detection under camera for social robot in smart homes. *Sensors, 18*(5), 1530.

2

INTRODUCTION TO IOT DATA ANALYTICS

Abstract

Internet-of-things (IoT) is an assemblage of sensors and actuators. The components work together by sensing and collecting information easier. They are self-contained, battery-powered, and are used in multiple ways. The machines aim by improving working performance, reducing price, and enhancing operational efficiency. The chapter discusses the architecture of such machines, different kinds of analytics used, application performance, use cases, and challenges of these devices. The future scope is the part of the reading.

2.1 Introduction to IoT data analytics

Investigating the data is the science or practice of managing examination to check the network. It is the method of deriving data and concluding from the outcome. The categorization of aptitudes permits us to discover the purpose of analysis and enables us to associate to the perception of IoT technology. The wisdom arrangement commences with gathering info at the station. They include events, pictures, and measurements. Erudition is described knowledge with meaning. Information is learning within a connection with combined conclusion and interest. The enlightenment understands with shrewdness. Each element of the education authority constitutes the past layer.

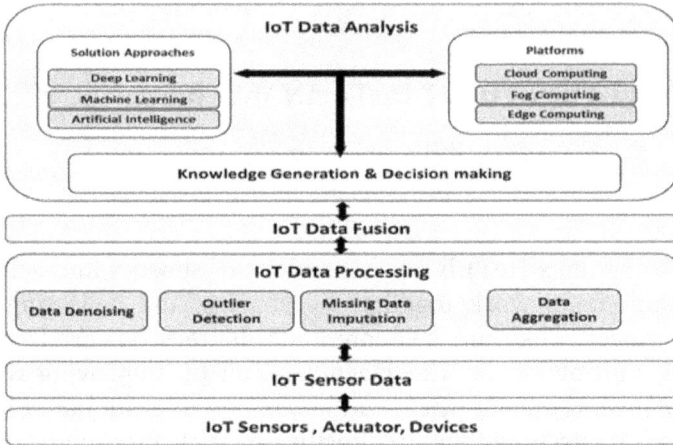

Figure 2.1 The basic architecture for IoT sensor data processing, data fusion and data analysis (Krishnamurthi, Kumar, Gopinathan, Nayyar, & Qureshi, 2020).

The IoT detector information tiers have different IoT sensing elements. These instruments estimate atmosphere and sense real-time surroundings modifications. It connects with the micro processing component, repository division, command segment, energy design, and unconnected transmission medium. The devices vary in dimensions, calibration capability, recollection, system capacity, and warehouse area. Wi-Fi, Zig Bee, Bluetooth, and LTE/4G mobile technologies are some of the transmission methodologies used by the devices. The information undergoes computation to extract indecisions for details investigation. The information fusion slab regulates different detector fact issues developed by various assorted machines. It focuses to create learning hub and aids in conclusion creation. Figure 2.1 portrays the architecture of IoT devices aiding in data analysis.

The IoT (Ambika, 2021) (Nagaraj, 2021)foundation contains IoT (Dey, Hassanien, Bhatt, Ashour, & Satapathy, 2018) connections. They are primarily IoT gadgets containing sensing devices and actuators. They are stationed at the corner of the system. The Fog devices help IoT machines in computation, managing warehouse, and developed interconnecting capacities. Storage joints have knowledge cores. They manage information collecting, computing learning examination, applying machine learning procedures, info distribution, etc. IoT employment practices the accumulated and investigated erudition to formulate assistance for the clients.

The framework (Marjani, et al., 2017) combines IoT and big knowledge analytics. The sensing layer has all the machines related to sensing. They connect using a radio system. RFID, Wi-Fi, ultra-wideband, ZigBee, and Bluetooth are different kinds of communication technologies. The IoT hub enables discussion of the cyber and interconnections. The topmost tier involves big learning analytics. It has a tremendous volume of learning obtained from machines. They are saved in storage and obtained through knowledge analytics utilization. These administrations hold API supervision and a dashboard to assist in the communication with the computation motor. It provides verdict assistance for complicated industries, structure supervision with the growth of evaluation practices, and the IT ecosystem.

2.2 IoT architecture for data analytics

Internet of objects is materializing as the item on the Internet. It predicts many materials or entities with diverse sorts of detectors and controllers. It connects to the Internet using multiple entry systems. It allows techniques to commute with each other. IoT is cyber techniques or a web of grids. With the enormous number of items and sensing elements coupled to the Internet, it is essential to gather accurate unprocessed information efficiently. The various contributors have suggested different architectures based on the framework.

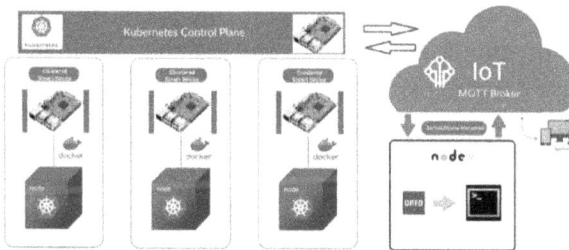

Figure 2.2 Technical solution for Labs of Things at UNED (LoT@UNED) (Pastor-Vargas, Tobarra, & Robles-Gómez, 2020).

The LoT@UNED podium (Pastor-Vargas, Tobarra, & Robles-Gómez, 2020) executes the edge tier using a collection of IoT devices. The instrument associates with the stock layer using the MQTT procedure. IBM stock contributor has the assistance of the stockpile-device stage. The detector stores its information in non-relational datasets. The warehouse is in the panel and aided conclusion tier. The design of the forum concentrates on the accessibility of low-cost Raspberry Pi machines. It associates with a grouping that facilitates the instrument link to cyberspace. It has two analytical collections enabling the linkage of new apparatuses. The initial cluster employs a distinctive frame for the administration and association of the instruments. It permits the collection of up to 40 Raspberry Pi machines, promoting the bond to the electricity net. This model utilizes the transmission as an essential component. The players give significance to the news. The system assigns the functions of the contributor, administrator, and agent. Endorsers document their welfare in certain transmissions using one or more backers. The peddler handles their appeal. It is liable for addressing the gush of messaging between editors and contributors. The publisher creates a text. The system transmits the message to the demanders. Figure 2.2 depicts the same.

2.3 Types of data analytics

2.3.1 Real-time analytics

Real-time IoT investigation (Verma, Kawamoto, Fadlullah, Nishiyama, & Kato, 2017) (Das, Dey, & Balas, 2019) is the procedure used to deliver balanced instrument assistance. It encloses management education, execution improvement of device derivatives, creative machine corporation benefits by exploring the organized IoT information employing corresponding repositories. The sources include web, estimation, and warehouse.

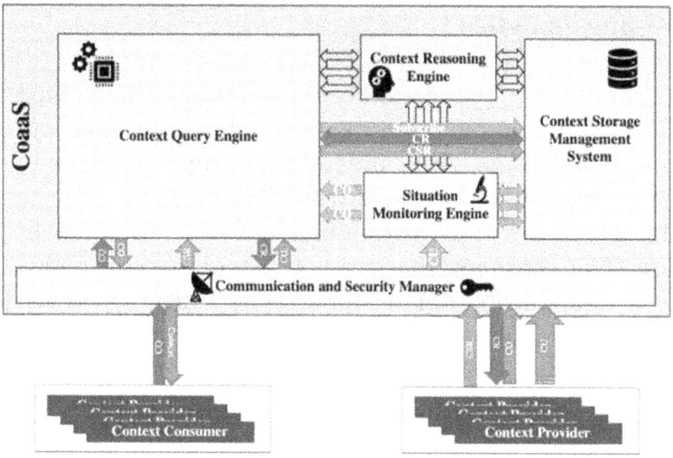

Figure 2.3 CoaaS Blueprint Architecture (Hassani, et al., 2019).

The CoaaS forum (Hassani, et al., 2019) has five associates. The transmission administrator is accountable for the preliminary conduct of received and sent dispatches. It behaves as a representative and disseminates coming data from Context providers and context consumers to the interrelated elements. It connects to protection supervisor. This administrator reviews the facts of arriving information and certifies submissions. It inspects whether the customer has admitted to the demanded assistance. It is liable for observing all the entering knowledge to recognize any questionable practices. The Context Questioning motor is accountable for scanning the arriving questions, developing and producing the questioning implementation strategy, and making the outcome. It brings needed information from workers on request. The Condition surveillance motor helps the constant supervision of coming text, investigate states from data, notice modifications in circumstances and deliver the notice. It surveys the real-time situation of the IoT commodities and explains their conditions. It creates the actuation system by reporting clients when required. The Context warehouse administration supplies characterizations of benefits and enables assistance finding. It stores contextual knowledge to provide question-answer duration and deals with issues like grid latencies and the possible unattainability of repositories. It accumulates and examines the documented context to promote self-alteration and efficiency balance. The Context Logic Machine figures the circumstances from unprocessed information. Figure 2.3 portrays the same.

2.3.2 Off-line analytics

Offline examination (Coetzee, Leeke, & Jarvis, 2014) supplies the capability to stack and investigate a sensed path without purchasing hardware. It functions with canned knowledge dispatched with the outcome or the details received from a purchase on the reasoning scrutinizer. It makes composition files with the reason investigation hardware. The software characteristics work accordingly. The offline study is applicable for technology without high necessities on answer duration. They execute examination by obtaining data from other sources. It logs into the venue using information purchase devices.

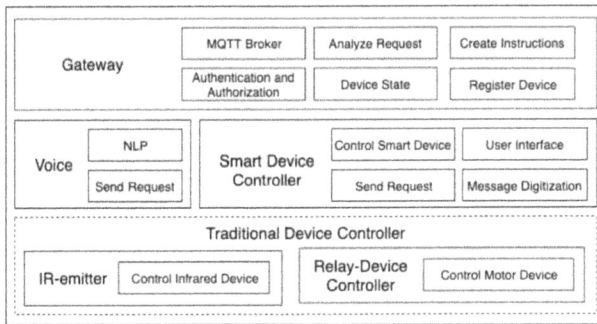

Figure 2.4 The proposed framework. (Wu & Huang, 2020).

The proposal *(Wu & Huang, 2020)* delivers a low-cost numerical conversion. The architecture has five components. The hub investigates the client's demand and provides acknowledgment. It decides whether the end machine requires storage assistance (when connected to the external system). It keeps the object's required information when offline and summons the stockpile. The device controllers broadcast customer appeal. The server examines the transmission to resolve the managing and announces rules. The customer gets the revised code of newly added devices. The voice process researches the intention described by the client's voice using natural vocabulary investigation. The voice component hears for audio intake, notices sound, and begins playing the audio to Lex. The Voice segment disseminates customer purpose to the hub. The system uses the information to analyze the intention. The server scrutinizes the teachings and posts them. A single-chip growth panel is employed to dispatch the command pedagogy. The component communicates to the IR-emitter. It, in turn, transmits unprocessed information approximating to the operations of the conventional machine to maintain the infrared instrument. The interior solenoid retains the existing loop to create a way and form a recent controller. The component establishes on the socket of the management appliance. The relay switch creates or demolishes the existing path and manages the swapping and stability of the electrical current to the widget after receiving a message from the gateway. Figure 2.4 represents the same.

The Electrocardiography alerts (Amri, Rizqyawan, & Turnip, 2016) examine the irregularities of the core. The features of Electrocardiography are easy, noninvasive, and moderately affordable. The characteristics are incredibly advantageous for fitness investigations. The sign recording receives alerts that can be examined and controlled. The lookouts compute to eradicate the noises generated by external elements. Some instances include the activity of the issue when the movement recording operation is in execution. The normal wavelets have unique qualities for noticing oddities in the cardiac approach. The electrodes of Electrocardiography have +, -, and relation electrodes. The electrode with + part establishes on the sick right hand, the – electrode nestled on the left hand and the earlier one, concern electrode element connected at the hip of the suffering. The asset technique for Electrocardiography surveillance performs on online topography based on the cyberspace of objects using mobile devices. The origin of Electrocardiography signs transmits the information using Bluetooth. The knowledge communicates to the stockpile or host through the cyber.

2.3.3 Memory-level analytics

The memory-level examination (Awan, Brorsson, Vlassov, & Ayguade, 2015) is applicable for data where the entire density is less than its memory group. The recollection of the existing gateway group crosses terabytes of storage. The system uses interior dataset methodology. The system is appropriate for real-time investigation.
The organization (Liou, Chen, Chen, & Lin, 2017) is balancing the program pile of the implementation conditions. The work focuses on memory function (trash cluster and memory administration). The collector has a general waste assemblage. The system marks and

sweeps in various angles. It depends on the aggregation methodology adopted. It belongs to the lowest epoch of entities (When an entity is assigned memory). The objects having descending years get better frequency than elevated epoch entities. The era of the object will elevate to the subsequent more heightened epoch (if the system does not reuse it). The means can decrease the weight of waste supply by reviewing subcomponents of the eras to bypass crossing all items on the cluster. Things in memory have their own source devices, defined by unique features. It will be put into the related origin (When the system assigns it). The performance has negligibly separated from the original waste grouping. The initial scrap supply mops items instantly after completing the pattern step. The rubbish assemblage cleans the commodity at the origin. The entities in the inventory illustrate the memory. The bases can be beginning junctures, from which the connected structure spans during trash accumulation.

The resolution (Kim & Park, 2015)has a co-invention assignment between SAP and the German Football organization. It delivers trainers and participants means to explore their own and competitor enactment. It results in an international programs dataset. The organization examined enormous quantities of participants' execution knowledge with the database. They studied videotape information from real-time cameras skilled of apprehending lots of details. It contains partaker part and pace. That facts are in SAP HANA that operates to authorize trainers to acknowledge undertaking measures for distinct performers. In Hoffenheim football activity headquarters, many workout structures and tools allow participants' interpretation. It conducts more useful and to scrutinizes their implementation. The technological associate institutions employ in the applications such as tape manipulation, 3D animation, spatial methodology, wireless system, and detector procedures. The specialized elements combine to make TSG 1899 Hoffenheim onto the Bundesliga.

2.3.4 Business intelligence analytics

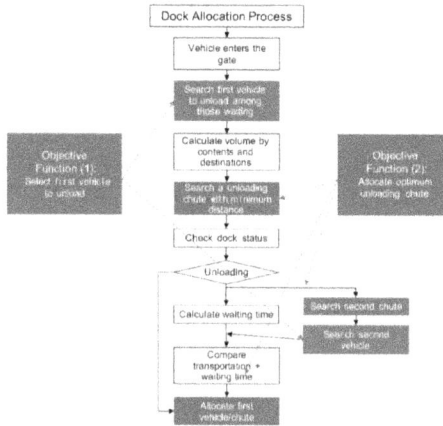

Figure 2.5Optimization of dock allocation process (Jin & Kim, 2018).

The gateway procedure (Jin & Kim, 2018) has three phases of messenger benefits. It includes picking the parcel, organizing it accordingly, and supplying it to the destination. The system has to avoid hindrance of the operations and hence conducted an investigation. The details are 240 million packets at the Daejeon server. The facts accumulated over three months between November 2016 and January 2017. This data develops comprehensive information on the acceptance ports at the gateway. It also uses paths, changes junction, shipment harborages, staying freight, and categorization executives. The data of quickest length between cargo and receiving piers, duration measurements, and automobile freight knowledge are combined with technologies. Figure 2.5 represents the same.

2.3.5 Massive analytics

Cyberspace of Objects is a system of enveloping devices. They link to Net using different detectors, automobiles, machines. It watches, witnesses, regulate the environment. The device entrenches commodities to perceive the background and intercommunicate with various objects. Massive investigation (Ediger, Jiang, Riedy, & Bader, 2010) is dissimilar from conservative knowledge controlling practices. It has a group of contests and prerequisites. It needs the computation to have a lower potential. The design is flexible with a self-load harmonizing capacity and good accessibility. It demands a constant repository for a brief period. It needs managing of the irregular pace of information, the inward learning having misplaced facts or uncertainties.

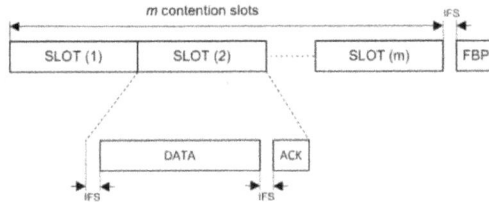

Figure 2.6 Frame structure of FSA-ACK and DFSA (Vázquez-Gallego, Tuset-Peiró, Alonso, & Alonso-Zarate, 2020).

The transmission (Vázquez-Gallego, Tuset-Peiró, Alonso, & Alonso-Zarate, 2020) has the end-machine and the grid interface that follows an organized framework in the learning Assemblage phase. It has packets with variance spaces. The end-instruments will have a single frame and haphazardly choose one of the places in each skeleton to ship the information to the web manager in the knowledge aggregation stage. The structures split into a designated number of assertion spaces. It approximates the time of a detail's sachet and a reply pack. It contains the watch duration between dispatch and accepts methods. It reimburses for beacon transmission, computation waits, and radio turn-around periods between ways. The packages can either collide or ignore sending or will be able to transmit parcels successfully in the case of *the FSA-ACK method*. The same is portrayed in figure 2.6. The packets are resent after collision in case of the FSA-FBP procedure.

IoT framework (Lv, Lou, Li, Singh, & Song, 2021) has three tiers. The perception stage acquires the information of connected items using advanced methodologies. The tools include detectors, radiofrequency labels, multiple knowledge aggregation, etc. The details transmit into digital signs. The network slab enfolds a portable transmission system. It balances the technical characteristics of the devices. The application tier integrates these methodologies with the fact's requirements of diverse enterprises. It performs a broad spectrum of wise technical explanations. It delivers low-price and high-grade answers for the endeavor. It assures data protection and executes industry prototypes. The system and transmission methods are fuzzy approaches of the machines. The entrance system and the backbone communication grid are essential in these procedures. The admission component has wide-location and short-length transmission methodologies. The backbone information grid depends on cyberspace procedures or transportable mobile methods.

The information (Saranya & Fatima, 2020) from IoT instruments collects and pre-calibrates for its fusion computation. A boundary-based clatter removal procedure details pre-computation which tries to tag unlabeled facts characteristics for precision in data fusion processes. Context-aware data fusion achieves integrating knowledge from numerous IoT appliances. This mixed information is understood using the Convolution fuzzy system for studying data fusion arrangements. Noises or pollutants in learning are removed by applying filters and rectifying them. It recognizes incorrectly marked details and reworks using category representatives where the standards work on uninfected databases. Cleansed details collections indicate knowledge that stays after filtering information. Self-Training Modifications happen by working filter on details groups. These filters work on the unclear info. Adaptive Noise Revision involves information clusters forming high-quality data set models to clear noises. These data sets perform K-fold cross-validations repeatedly. In each iterated round, predicted labels matching corresponding inferred labels results in high-quality knowledge. Dynamic Bayesian system separates facts into period pieces for describing conditions of an illustration where Hidden Markov Models are used to locate its noticeable manifestation. Convolution fuzzy system-based Forecast has three tiers - intake, convolution, and SoftMax stages. It decreases processing overheads and improves prognosis precision. SoftMax tier uses a SoftMax procedure to compute allotment options of an occurrence in various circumstances. The authors use the standard bloom filter and extends the same.

2.4 Analytics method

2.4.1 Classification

This is an investigation where the inputs are categorized based on the similar attributes they possess.

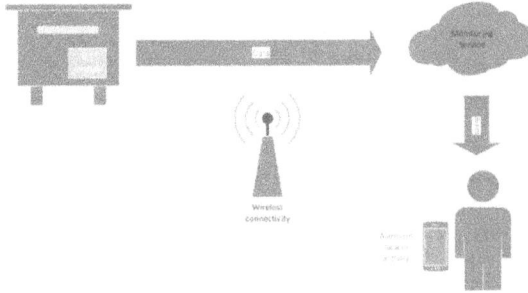

Figure 2.7 IoT-based bee swarm activity monitoring service (Zgank, 2020).

The study (Zgank, 2020) suggests two methodologies. The seized apiary acoustic information includes various bee workouts. The supervising explanation connects hive movement signifying resonances. They categorize them into routine and beehive activities. Open repository Apiaries Scheme enables ICT methodology in the discipline. The system uses different single panel systems and mics in the initial phase to gather information. The detectors accumulate ecological knowledge in the next stage. The work stores grabbed learning on the stockpile. Figure 2.7 illustrates the same.

2.4.2 Clustering

The procedure also categorizes the intake having labeled set of attributes.

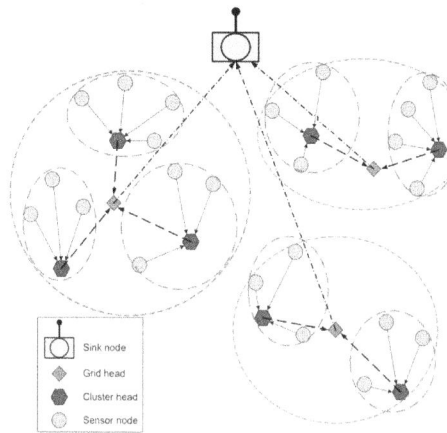

Figure 2.8 Three-layer clustering topology (Faizan Ullah, Imtiaz, & Maqbool, 2019).

The atmosphere (Faizan Ullah, Imtiaz, & Maqbool, 2019) observes, feels items, be recognized, and achieves any pre-specified activity. Customers access them using cyberspace, obtain identification and take measures to maintain the domain. IoT media supply altered qualifications in the enterprise. The industrial future supplies easiness for the idea-bringers and implementers. It guides to more profitable judgment-making. Recent IoT podia need to communicate and deliver some benefits and characteristics to promote their selection. It supplies many assistances and technologies to information purchase and investigation, instrument administration, combination, protection, understanding to buyers on the procedures, and proficiency to recognize and control appliances. Various IoT prototypes work on identical hypotheses or institutions and mediums as a benefit. Figure 2.8 illustrates the same.

The suggestion (Yao, Wang, Shen, Kong, & Ning, 2019) makes two amendments compared to the previous doings. The first selects the start grouping bases following the length and the knowledge thickness to overpower the regional balance and 490 contracts outstanding collection outcome. Assuming that the deviation or likeness is the fundamental measure in forming groups, the estimation by space, possibility viscosity, the work utilizes the moderate stretch by merging with the intermediate details thickness to decide the preliminary gathering focusing nominees, to avoid the boundaries of noise. The mean length is the middle value of all the spaces among the representative slots. The second operates the lowest inter-cluster randomness rather than seeing the imperfect bunch to enhance the grouping precision.

The recommendation (Puschmann, Barnaghi, & Tafazolli, 2016) specifies intake information. In the k-means procedure with arbitrary resume joins, the groups are designated. The consequent collections reclaim to organize into equivalent facts groups. The collecting means adjusts the centroids of the sets based on the present allotment in the information sequence. The details float discovery is initiated by evolutions in the numerical particulars of the likelihood thickness procedure. The rationale has attributes of stochastic convergence. Conjunction in a mean square indicates junction in possibility, suggests intersection in odds and allocation. During workouts, the system keeps the traditional variation and envisioned weight of the knowledge with the existing allotment. When computing new input learning, the system finds the trail of the anticipated text and expected divergence modifications giving the new significances.

It is an enhanced subspace grouping procedure (Cui, Jing, Zhao, Zhang, & Chen, 2021) having a post-procedure approach. Its pacts with hyperspectral knowledge collection (domain of IoT). The technique is a post-methodology technique. It hovers spareness and links of the replica measurement matrix in enhanced subspace batching. The work defines close neighbors. It does not assume mutual next-door numbers. It emphasized the associations. The system determines immediate adjoining ones from a representative's next-door ones. The work uses the non-ruled arranging procedure. The recommendation conduct investigations on the standard graphics databases. HSIs collection demonstrates the significance. It is ominous of the post-procedure method. The recommendation divides the examination into two methods. SSC is approximated with SSC-PSCN to demonstrate the usefulness of PSCN from various viewpoints. COIL-20 is a bunch of images of 20 topics. Cameras take photographs of each issue from 72 distinct inclinations. The AVIRIS detector obtained the Indian Pines location. The number of spectral rounds is 224 and has 145×145 pixels.

2.4.3 Association rule

The IoT contains processing knowledge and technical approach by offering the predominant position to the items and logic examination essentials. The Internet of Things integrates the real-world to the forthcoming track of mortals with digital software to sufficiently react to the activity when observed precisely with augmented dispersed authority on the enterprise.

The design (Aljawarneh, Vangipuram, Puligadda, & Vinjamuri, 2017) is an IoT framework. It includes documenting the developed time sequence of dynamic information connected with the host. The knowledge created is time sequence and held at time intermissions. It is needed for investigation and examining the tendencies. This time succession is not appropriate for researching the active directions or for recovering the current practices, seasonal routines, reducing customs by employing the (existent) procedures and approaches in the writings. This needs a pre-computation of the registered time sequence. The output of the step is a time-stamped temporal dataset extending (horizontally) when approximated to a conventional Relational database management system. The procedure evaluates practice leaps and distinction grades (between routines using vague gaussian disparity calculation).

The work (Ozawa, Ban, Hashimoto, Nakazato, & Shimamura, 2020) starts with emanating the collection of recurring objects and their support totals. Erase all things from the commerce not fulfilling the lowest support restraint. The regular articles in a title plateau are in the dropping ranking of their commonness. The next phase creates an FP-tree by injecting samples into a structure with a seed. The objects in marketing organizations are in the same hierarchy as in the heading plateau. All devices directing to the same thing are jolted in the list and measured by traveling this index. The title components connect to the related objects. Repetitive digging of the FP tree can produce enormous article collections without yielding nominee objects and sampling them. The procedure initiates from the base, making the dependent object floor. It has a collection of prefix routes in the FP-tree recurring with the suffix object. The dependent FP-tree with calculations focuses on the actual structure resembling a collection of samples with each device obtaining kid total. Recurring maturation finishes when no particular objects on the feature satisfy the lowest support benchmark. The precomputation resumes on the remaining header objects of the initial FP-tree. Once the recursive procedure completes, the system finds all large itemsets that fulfill the lowest support restraint. The connection practice innovation starts.

2.5 Use Cases

2.5.1 Smart metering

An intelligent measure is a machine that electronically documents the usage of electrical strength information among the measuring device and the administration method. Gathering and investigating clever measure info in the IoT ecosystem support the decision-maker in foretelling current expenditure. The analytics is estimate requirements to limit trauma and repay diplomatic purposes in particular pricing policies.

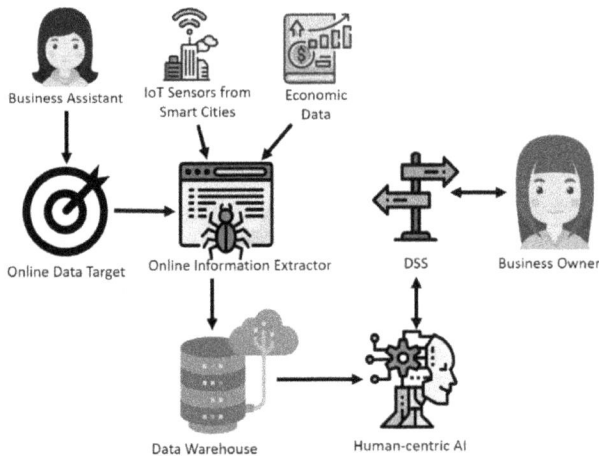

Figure 2.9 Overview of the system for supporting real-time business decisions (García-Magariño, Nasralla, & Nazir, 2020).

The company buyer (García-Magariño, Nasralla, & Nazir, 2020) determines the scope to finance. The professional associate examines the most reputable portals having real-time information. They choose online knowledge. They may be websites and the significant learning for recovering the information. Then the online data extractor is configured with this knowledge. They accept facts from the online results and feed them into the information repository. Diverse investor approaches achieve the DSS for supporting the company proprietor in their judgment. The practice creates Human-oriented simulated mentality solutions based on the incremental computation of characteristics or real-time knowledge. Amidst collecting information and estimates, the operation extricates the time range of inducing averages. They are medium amounts estimated from a definite number of current examples for every minute. It analyzes them with another period range used in the education stage. The Simplistic investor procedure concentrates on compact interval properties based on prevailing cost. These tactics buy Bitcoins when the value is more economical and trade them when the rate is more crucial. It examines a specific merest outset of diversity to evade lack of capital due to the charges of regular exchanges of money. The policy depends on the course of average means. It increases regarding specific periods and cost outsets. 10,000 min in arbitrary interludes with numerous policies were recognized. Simulations have interim sizes from 5000 min to 100,000 min with levels of 5000 min. figure 2.9 illustrates the same.

The planning (Liu & Nielsen, 2015) splits the system into three zones. Each course serves as a separate practice to engage the overall demand of streamlining intelligent measures and investigating the knowledge. The processing tier manages the ingested info in various learning calculation operations. The course takes care of the records. It consists of divergent computing tasks. These execute in the corresponding underlying data processing systems. All the flow of work is registered to be performed either once or recursively at a definite interval. A course has various associated outcomes. They practice a distinct computation assignment and administer on an underlying info calculation method. The analyzing tier has a high-performance store, associated archives, interconnection utilization, and description generator. They collectively assist clients in analyzing inquiries. This method is a web-based reinforcement of practicing Tomcat as the administration host.

The architecture (Hu & Tang, 2020) includes three main components: the cloud center, the communication network, and the EI-smart meter. The cloud center has numerous applications in data processing and analysis. It is responsible for processing and analyzing data uploaded from all smart meters. It uses three components. The house power machines interface with the EI-smart measure over transmission technologies utilizing the house region system. The multi-cell coverage assembles the data uploaded by the devices. This information is then conveyed to the repository hub using the WAN and the neighborhood area network. All EI-smart measure identifies valuable knowledge from home strength mechanisms. It then builds secure connection joints and accumulates the learning from the machines. EI-smart measurement instrument examines the data obtained by the detecting zone. It conveys the outcomes to similar tools. They store by analyzing the adaptability of another purpose. EI-smart measure offers facts digging and interpretation of the household power data for utilization. EI-smart instruments contribute many creative administrations such as weight review, duty forecasting, and capacity administration.

2.5.2 Smart transportation

The transit arrangement is an IoT-based application that intends to establish the metropolis idea. The shipping method deploys sturdy and exceptional intelligence technologies for the administration of intelligent towns. Satellite exploration and sensing applications in vehicles, vessels, and aircraft help the tracking of objects. The routing in vehicles optimizes by practicing the volume of usable unrestricted info, such as transportation difficulties, street situations, shipment locations, climate statuses, and refilling sites.

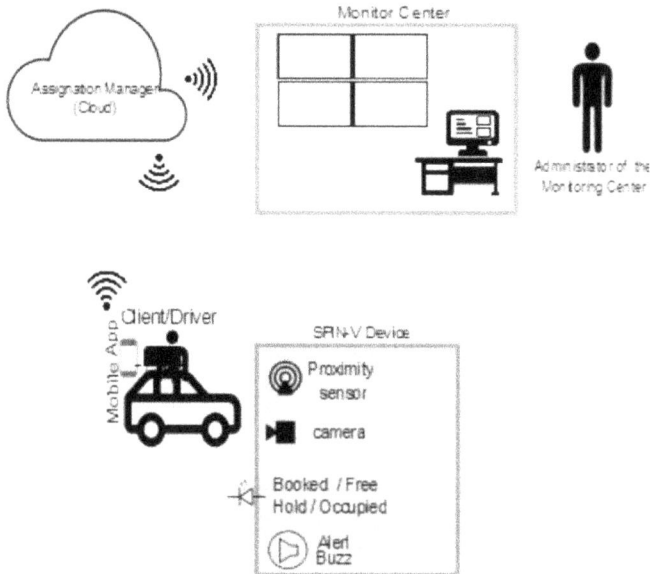

Figure 2.10 Network architecture of the SEI-UVM (Luque-Vega, Michel-Torres, Lopez-Neri, Carlos-Mancilla, & González-Jiménez, 2020).

The suggested SEI-UVM (Luque-Vega, Michel-Torres, Lopez-Neri, Carlos-Mancilla, & González-Jiménez, 2020) is an intelligent parking scheme. The system uses the parking design. It has three components. Each parking area has a SPIN-V discovered in the center of the backend of parking slots. The dynamic system is unrestricted for the clients to organize and prearrange a parking area. It also scrutinizes the center. The owner of the private parking areas manages and handles the parking slots and the registrations. The knowledge communicated by the framework is kept in a dataset and monitored by OBNiSE. All the data transmits from the SPIN-V to the stockpile. The customer can add his details and record the automobile's authorization description. It can scan the parking slots, book an area, and acquire various path suggestions. The SPIN-V has a distance detector to notice the admission of an automobile to the parking area, a camera to capture a picture of the car's plating, and an LED arrow and buzzer to report the situation of the parking area. Figure 2.10 portrays the same.

2.5.3 Smart supply chains

The achieved knowledge is used in onsite and off-site technicians run diagnostics and rehabilitation opportunities to make relevant judgments. The outcome is enhanced engine uptime and consumer assistance. Installed detector practices interacts bidirectional and present distant convenience. Essential practices employed by in-transit distinctness are RFIDs and store-based Global Positioning Mode. It renders position, connections, and different tracking knowledge. This information will be the spine of the stock series sponsored by IoT methodologies. The learning inferred by the equipment will present complete clarity of an object transmitted from a producer to a middle-man. Information gathered via RFID and GPS methods will enable equipment connection administrators to improve electronic purchase and detailed distribution knowledge by foretelling the moment of shipment.

2.5.4 Smart agriculture

Sensing devices are the players in the intelligent horticulture. They are situated in courses to collect data on moisture tier of dirt, container width of flowers, microclimate situation, and moisture level, and estimate climate. Instruments communicate received info using system and transmission tools. These knowledge passes through an IoT hubs and the cyber to end in the analyzing tier. The investigating panel prepares the data obtained from the sensor network to issue commands. Automatic climate control according to harvesting requirements, timely and controlled irrigation, and humidity control for fungus prevention are examples of actions performed based on big data analytics recommendations.

2.5.5 Smart grid

Watching communication cables, creation components, substation phase, clever measurements, and knowledge aggregation from households creates an enormous dataset from the system. It keeps the wisdom in stockpiles for investigation. Storage backed with an IoT network manages the information. The datasets require more enhanced techniques to seize learning. It has to remove noise, organize data, and examine. The learning advances as the methodologies move. These characteristics correspond with the facts IoT machines development, and thus the information rendered in the system. IoT incorporated methods form a complicated corresponding tangle. It has a considerable magnitude of info stowed in warehouses. The violation of information protection is a grave problem.

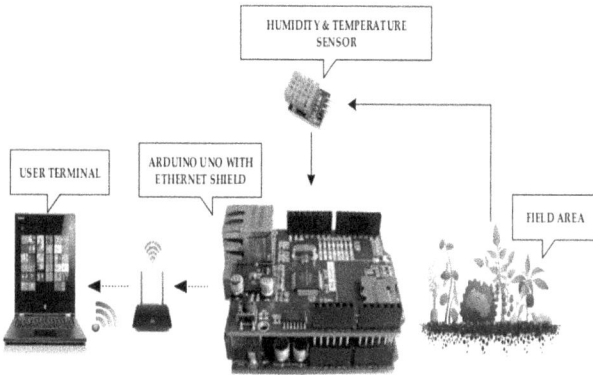

Figure 2.11 Design of the weather monitoring system for smart farming. (Chhaya, Sharma, Kumar, & Bhagwatikar, 2018).

The architecture (Chhaya, Sharma, Kumar, & Bhagwatikar, 2018) supports regional and global systems. The approach uses climate constraints. The DHT11 detector senses the climate and wetness of land. A thermistor and capacitive device are used. The system uses Pin 2 of Arduino Uno to intake readings. It has an HTML webpage to monitor the climate. The customer can access the webpage. The active keeper formatting procedure allows a host to designate an IP address to corresponding consumers. Ethernet protection is used for the Arduino web server application. Figure 2.11 portrays the same.

2.5.6 Smart healthcare

Wellness data Interchange (Zeadally, Siddiqui, Baig, & Ibrahim, 2019) (Sakr & Elgammal, 2016) improves healthcare distribution by implementing the capability to electronically administer healthcare administrations reliably and securely. The client-mediated switch gives victims passage to their computerized documents. It allows them to trace their fitness limitations. It resolves incorrect billing of pharmaceutical information and renews them. The addressed transaction handles when a healthcare system carries significant learning

such as lab analysis outcomes and remedy dosage to different professionals included in the interest of the corresponding victim. The
query-based trade usually happens in unplanned corrective concerns when a healthcare industry requires the prior fitness reports
of a new outpatient. It demands entrance to recordings through the
arrangement.

2.6 Applications

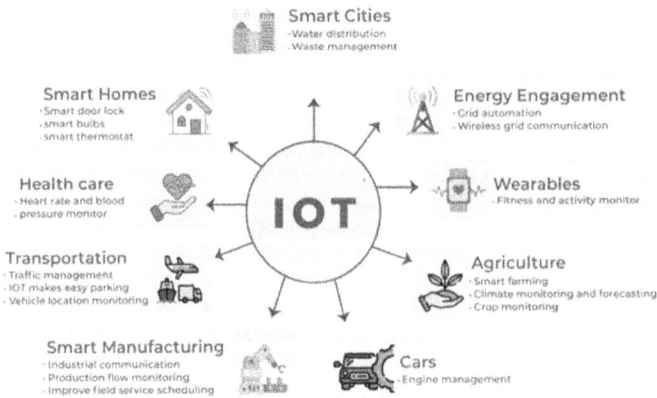

Figure 2.12 IoT applications and services. (Alghofaili & Rassam, 2022).

2.6.1 Smart cities

IoT information investigation is a key to obtaining and imparting
wise decisions to metropolitans. The virtually infinite processing
and warehouse are to be maintained by stockpiles. IoT architecture
and evolving knowledge from municipality appliances require quick
answers. Some of the applications include shared security and crisis
reaction. The scenario requires high bandwidth usage and low
communication cost.

The framework (He, et al., 2017) is multiple phases fogs processing
structure. The system includes dynamic nodes with disseminated
repositories and reliable devices. The machines broadcast unprac-
ticed warehouses activities to deliver detailed examination in appliances and portable devices. They can partake in a tree-structured

stockpile. It operates with conventional isolated storage locations. The instruments are opportunistic and trustworthy to alleviate the issue of massive frame acquisition. The employees connect to these machines using the request from their heads. They are accountable for transferring their processing storage, essaying assignments, surveillance, and documenting functional and transmission depositories to the captain. The device has a moderator. Numerous experts can be a part of the system. This methodology enhances the trust of the system. The authorities may physically position themselves with their subordinates or individually. The aces have the primary duties to make, administer, and organize the assignment. The investigation executes this system with the aid of disseminated motors. The experiment has a collection of processing warehouses with one computer and 8 Raspberry Pi 3 credit card-sized microcomputers. The Raspberry P links to a WiFi ad hoc network through their built-in wireless 802.11 model. The appliances nestle on the system and assign 700 MB RAM. The framework uses Spark.

2.6.2 Retail and logistics

Eco-driving (Hopkins & Hawking, 2018) is a system developed by BDA at organization. It links to decreases in combustible expenditure and CO_2 emanations. It brings both financial and habitat benefits. Vehicle telematics used enhanced driving behaviors. Distant command stations control existing sensor knowledge from the business. It captures the data concerning pace, position, braking, and motor info. It notifies prospective education applications. A mixture of vehicle telematics and geospatial data facilitate proactive signals assigned to motorists. The data concerns potential future dangers. Camera-based methods enhance operator protection and exhaustion administration, seizing indications of significant driving transactions and collecting learning to the storage. BDA increases vehicle routing, advises optimal combustible procuring events and positions, and anticipates imminent and proactive preservation programs. Changes in utilization and routing have the potential to lessen transportation bottlenecks. It is accountable for damages in potency, improvements in combustible loss, air contamination, and sound. It can induce anxiety, aggressiveness, violence, and threatening reactions in motorists. The ominous investigation produces refueling and subsistence agendas. It has the potential to be utilized by all transportation companies and create modifications in client am-

munition prices. It improves the routine, effectiveness, and existence anticipation of potential vehicles.

2.6.3 Healthcare

Infirmaries are increasing to recognize and minister the sick. To
keep processes, clinics frequently depend on progressive methods.
The expansion has wise automobiles are merged with devices. It
permits in-ambulance analysis, helps therapeutic executives to position suitable therapy before the victim reaches the clinic. Intelligent
vehicles would need trustworthy diagnostic medical detectors, protected transmission connection with the infirmary. It assists clinics
in working spaces. The operation is aggressive by employing swap
communication and producing movements among executives, doctors, and therapeutic machines in the operational chamber.

The semantic web prototype (Reda, Piccinini, & Carbonaro, 2018) is
a layered strategy. Each tier furnishes a collection of distinctive
roles. The top of the heap contains ontology vocabularies, practice
speeches, question terminologies, reasoning, sense tools, and confi-
dence. Ontologies comprise the spine of the framework. It describes
ideas and connections of a discipline and identifies sophisticated
restrictions on the sorts of repositories and their attributes. Convention tongues permit registering presumption regulations in a stand-
ardized method. It employs sense using specialization. There are
reason and sense on high tiers that supply the visionary reinforcement needed for logic and speculation. First-order reasoning and
explanation help the logic approach. It makes assumptions and pulls
new understandings based on the warehouse text depending on ontologies.

The device (Kodali, Swamy, & Lakshmi, 2015) eradicates the requirement for a fitness maintenance expert at periodic gaps. It furnishes omnipresent surveillance techniques operating detectors,
hubs, and stockpiles. It examines and keeps the knowledge and
transmits it wirelessly to doctors for additional investigation. A
medic can gain entry to the suffering information employing cyber
facilitated instruments. The system diagnosis it and can specify suitable therapeutic administration. Intel Galileo Generation 2 panel
works as a hub to operate the acquired learning. It handles an online
host to transmit the knowledge to the warehouse. It has a 400MHz,

32-bit Pentium architecture-based single-center, single-threaded Intel Quark microprocessor, 7-15 volts battery, and 256 MB of RAM. The work (Lv & Qiao, Analysis of healthcare big data., 2020) is designed with 100 inquires. The study distributes the questions among the medics, nursemaids, managers, and mechanics. Out of 92 queries, 89 are reasonable. The recommendation outlines protection hazard characteristics of fitness learning. The questions have four elements- information cluster, warehouse, methods used to visualize data, and collapse. The exact text of the queries contains therapy method, confidential grouping by organizations, adversary team, exterior adversarial invasions, framework devastation, vague appropriate regulations, insufficient uniqueness authorization, unfinished disorder analysis procedure, inadequate admission approach, preliminary safety standards, information encoding, ineffective secret understanding, poor regulatory calculations, knowledge investigation, essential misplacement, and details backup.

The suggested Design (Bhoi, et al., 2018) operates Accelerometer and Gyroscope detectors for developing and managing the information. An accelerometer is an inertial active sensing element. It is qualified for sensing acceleration. The gyroscope adjusts the direction of the mortal. The analysis has four conditions (sleep, sit, walk and fall). The database used is MongoDB. Knowledge precomputation happens before the investigation of collected details. It notices the misplaced text from the database. It concerns the expected precision of the category procedure. The work has documented fact and category tags. The titles have the volume of the accelerometer and gyroscope detectors. Two processes in Python instruct the prototype by assuming the data. The knowledge methods discover the invisible practices. It maps the intake details to target variety. It delivers a classifier standard to organize the more recent or forthcoming knowledge. The representative categorizes the activity. The warning method will automatically call respective using a Python model.

Five associates (Jeong, Han, & You, 2016) perform fitness design. The procedure begins with the detector gathering information. The assembled knowledge goes to an intellectual grid to disseminate with the method. Transmitting learning through the system is held at stockpile. The details undergo investigation for proper decision in the fourth component in big-data analysis. Filtered information undergoes transmission to a wise infirmary to announce the outcome to fitness experts. The system takes steps for findings and therapy.

The primary IoT framework (Rahmani, Babaei, & Souri, 2021) has enterprise, appliance, middleware, grid, and perception. The textual tier is accountable for information investment from different detectors in the system. It maps onto the perception and web slab. The event tier focuses on the occurring circumstances based on a specified regulation. It suggests the same to the middleware slab. Diverse benefits instructed in the service slab are pieces of the application and business tier. The work has two decks. The first tier is liable for obtaining real-time information. It is trustworthiness having characteristics of methods. The communication waits estimates transmitting and accepting packages in outdoor conditions at different duration gaps. The second tier has logic and decision-making. It has predefined constraints allocated to the NESPER motor. It uses complex event processing tools.

The detector (Muthu, et al., 2020)is employed to witness and forecast disorders in mortals. The sufferer details organize into demographic information, attributes of the cancer analysis, core illnesses, diabetes, blood stress, etc. The specified characteristics react to physiological gestures, body motions, and organic essences. Knowledge treats them as a series of subsequent actions like learning precomputing and info investigation. The original sign from the noisy sign retrieves in the first phase. The work examines frequency and duration territory in the beacon. Attribute extraction completes extracted the characteristic by employing balancing measure. The details of infections accept the well-determined illnesses datasets. It gets a more examination of findings. The logic-based cleverness Management executes a neuro-ambiguous approach. The stage concerned devises forecasting the colorectal tumor, repositories for information digging, and document accounting. The work precomputes knowledge gathered from different origins. It eradicates the monotonies and inconsistencies. They assure the grade of information. The computation and category of knowledge assemble about the sick using the framework.

2.7 Challenges

2.7.1 Privacy

This development has many advantages by amending the way people live to carry out day-to-day jobs and potentially altering the globe. The system evolves where clients require better protection, which decodes into confidentiality. The massive quantity of information generated is overwhelming. It makes more access points for pirates and leaves acute details helpless.

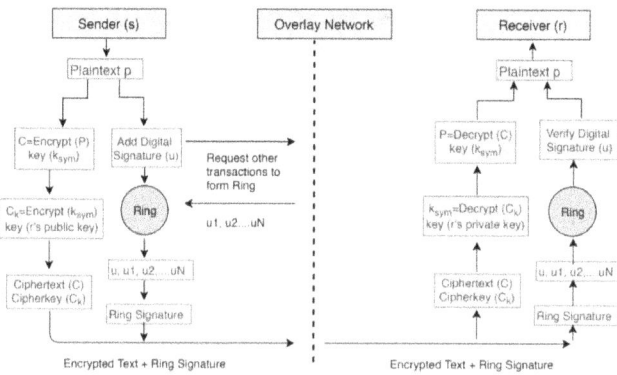

Figure 2.13 Block diagram of the framework (Dwivedi, Srivastava, Dhar, & Singh, 2019).

The authors (Dwivedi, Srivastava, Dhar, & Singh, 2019) suggested the idea of a data management system and its confidentiality. It instructs a precomputation procedure before conducting information mining. It bypasses details leakage and useful knowledge misplacement. The client information, accumulated by wearable detectors, wellness surveillance appliances, and mobile, is sensitive and secret. Immediate investigation offers hazards of exposing sick workout conduct, place, and other fitness-irrelevant confidential data. Figure 2.13 portrays the same.

2.7.2 Data mining

The hybrid IoT territories yield a tremendous quantity of facts. This information is originally unprocessed that requires to be computed.

Different algorithms are used to mine the datasets. This section details the same.

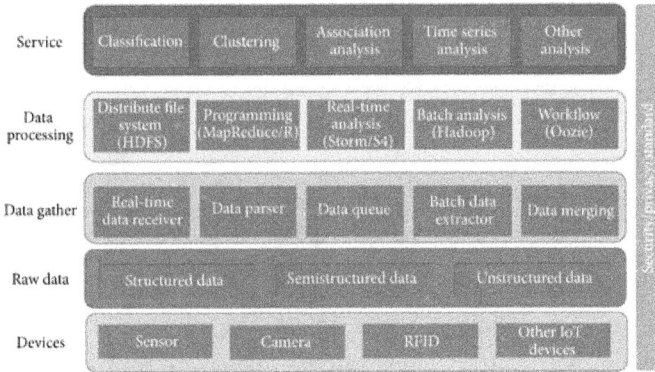

Figure 2.14 Data mining system (Chen, et al., 2015).

The strategy framework for Internet of objects and knowledge digging management (Chen, et al., 2015) uses six stages. Enormous sensing machines combine to apperceive the earth and develop information constantly. The system comprises organized knowledge, semi-designed learning, and formless education. Real-time wisdom and collection dataset are explored, investigated, and combined. The data combines using Hadoop, HDFS, Storm, and Oozie. The system assists in using these tools. Protection, confidentiality, and benchmark are essential in the framework. The approach shields the knowledge from illegal authorization and information exposure. The procedure creates knowledge aggregation, sharing, and excavating for future use. Figure 2.14 portrays the same.

2.7.3 Visualization

Data presentation methods (Chen, Guo, & Wang, 2015) connect to the visible investigation. It seeks to promote the studies of web managers in the methodology of surveillance and supporting grid fitness. The measure database and measurements have little help for collaborating inquiry. The traffic picture concentrates on the high-volume traffic parts.

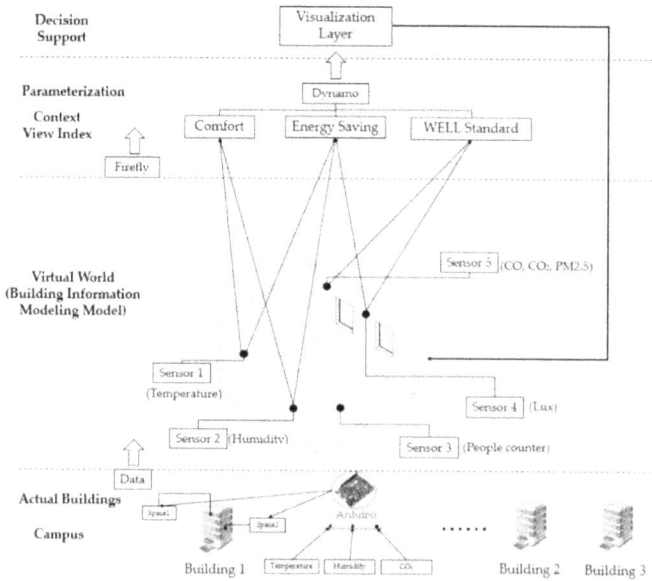

Figure 2.15 Visualization framework (Chang, Dzeng, & Wu, 2018).

The work (Chang, Dzeng, & Wu, 2018) has four segments. It contains detector information aggregation, knowledge integration, conditional administration, and presentation designs. It employs Dynamo atmosphere to install an Arduino association. It creates a detailed representation procedure. It generates charts and the 3D outcomes in the workplace. It uses Python programming to work forecasted Poll investigations using customized devices. This investigation analyzes scholar classes in the teaching space. It considers their convenience influenced by the climate and air moisturizer at 80–130 cm beyond the foundation. The suggestion embeds detectors in nine places along the borders and in the middle and junctions of the site study setup. Figure 2.15 depicts the same.

The investigation (Shao, Yang, Juneja, & GSeetharam, 2022) is an idea trial to identify the necessary conquest measurements of demand of information study. An IoT-based knowledge illustration architecture overwhelms the weaknesses and improves the enterprise intellect arrangement. Detectors and learning transcribers are tools and devices in the production procedure. They deliver extensive magnitudes of information to make a new path. The study of accomplishment estimates the performance by employing a knowledgeable representative. The topics receive training on how to utilize stationary equilibria. The industry revenue increases by using the random examination. Learning illumination makes attendants evolve more elegant. Real-time information wisdom and contextualized facts emanate from the knowledge gathered by different detectors. The procedure affects a community's judgment-making operation (using collaborating graphical presentations of info). A storage processing combines continuously and the interrelationship in comprehensive knowledge computation and reviews. The element examines the preliminaries of knowledge examination, a mix of learning and Yield distribution. The enormous investigation is collection, interexchange, and removed data of available facts. It authenticates practices, understands from data received. Deep knowledge enables approaches to comprehend and analyze extensive collections by converting them. It removes the complex intake and transforms it into comparatively more uncomplicated concepts illustrated in the earlier deck.

The AGENT knowledge administration (Quan, Yang, & Luo, 2021) ought to comprehend the exchange among the original information and the system. It is a medium for assemblage and control. It has two components. The representative division of the client assembles multi-tier and multiple directional information. It contains procedures as a database, kinds of facts, and stockpile information(grade). The customer's broker division can describe linked grouping constraints (if necessary). The initial details group representative is a strategy for the segment posted by the consumer and the knowledge cluster mediator to convey with the assemblage agency and interaction representative details among systems. The numerous knowledge relations can be made simultaneously (When the agent coordinator communicates). The mine describes information according to the essentials of the details managing component. The removal procedure has two factors. Pull-out details based on information changed. The international digging ministers the whole grid as an assemblage and executes moves. Regional digging is the methodology of dragging precise text or procedures based on hierarchy. The details computing component will explore for the identical knowledge as the extraction yield and transfer, excavation, and change equivalent info (after pulling the proxy information).

2.7.4 Integration

The Cyberspace of objects model involves clever and self-composition devices connected in a vibrant and international grid organization. It symbolizes the disorderly methodologies, allowing omnipresent and prevalent calibration procedures. IoT has small items, widely spread, with inflexible warehouse and computation capability. It focuses on trustworthiness, execution, protection, and confidentiality. These devices combine with other technologies to make architecture work better.

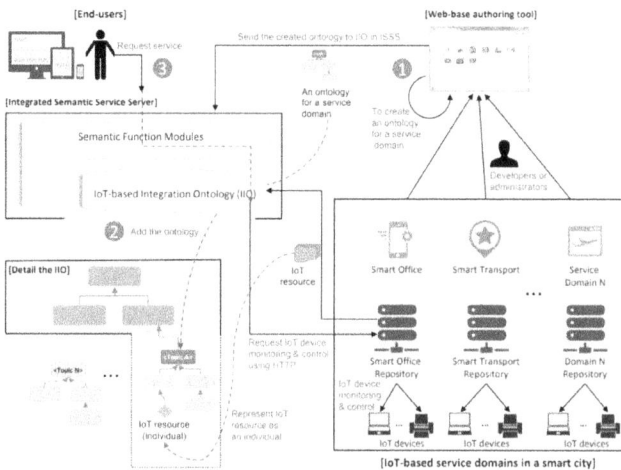

Figure 2.16 The overview of the integrated semantic service server (Ryu, Kim, & Yun, 2015).

The system (Ryu, Kim, & Yun, 2015) has two components. The web-based authentication device makes an ontology employing an online browser to update assistance field learning. The element has three sub-components. The assistance territory subject area permits designers or managers to construct the class subject. The ontology mechanism and association domain intake the many information like title of the course, entity attributes, information characteristics, field, scope, and limitation to make an ontology contemplating the assistance territory learning. The relation domain infuses facts about the device warehouse. The Reasoning Assistance gateway has five members. The logic signifier functions as a semantic encoding to portray text obtained from the web-based authorization device and exterior IoT benefit medium. The ontology registering adds the web ontology language and resource description framework acquired from the semantic explanation element into the IoT-based service integration ontology. The semantic discovering component functions semantic findings using SPARQL. It constructs questions for semantic conclusions in the machine-based service integration ontology. The assistance connector develops controls for cooperating with exterior device media employing the individuals obtained from the semantic explorer. The instrument-based service integration ontology manager is accountable for supervising. It reworks the device-based service integration ontology. Figure 2.16 depicts the same.

The Cyber of objects suggests a sequence of interconnected populations, entities, appliances, and information over cyberspace for management, assistance medium, and administration. The progressive benefits need a tool to gather, research, and compute unprocessed knowledge from detectors. Datasets created and executed to operate with them have distinctive necessities and features. The requirement of stockpile podia reduces the chances to join this auditorium and take benefit of its accomplishments and assistance for many companies of varied measurements. The databases operate in learning and receive from diverse sensing elements. These details are developed and computed in a multiple-stage prototype and advance to the judgment-making podium. These measures include information creation, assemblage, association, calibration, and judgment creation.

The system (Doan, Kayes, Rahayu,, & Nguyen, 2020) is an active catalog-based architecture for sequencing information from numerous IoT repositories. It includes two principal subsidies. The range of questions is duration-based necessities. Their answers explore the directory of the condensed knowledge in the warehouse. The example can remove learning from numerous references constantly using the windowpane method. It describes every collection of computing information with a measure equivalent to a window height. This stage uses the earlier creation on window removal. The location of details is in our period catalog. The facts design is in couples. The credentials are datasets and keep all characteristics of each document. The floating-point consolidated information can be squeezed by involving a lossless reduction. It enhances the compaction balance to improve the depository capacity. The duration directory is purified to explore the catalog based on the period and client's questionings.

2.8 Open Issues

The massive use of the Internet nowadays has led to the enhancement of new methodologies and appliances. Internet of Objects is the adoption of the number of interconnected machines. It is increasing at an exponential pace. The technology has also let to collect enormous facts magnitudes. The IoT exploration defines the term as the investigation of components of the enormous IoT knowledge to receive the industry's worth and navigate wise judgments. The tremendous IoT learning computation and research of unsubstantial magnitude without considering the history of facts. It seeks to acquire observable understandings from the mining of documented information. It classifies into an explanatory and diagnostic investigation which furnishes visualization or reports-based statistics of the IoT strategy and malfunction warning of the detectors and tools.

References

Aamir, M., Masroor, S., Ali, Z. A., & Ting, B. T. (2019). Sustainable framework for smart transportation system: a case study of karachi. *Wireless Personal Communications, 106*(1), 27-40.

Abdel-Basset, M., Manogaran, G., & Mohamed, M. (2018). Internet of Things (IoT) and its impact on supply chain: A framework for building smart, secure and efficient systems. *Future Generation Computer Systems, 86*, 614-628.

Alghofaili, Y., & Rassam, M. (2022). A Trust Management Model for IoT Devices and Services Based on the Multi-Criteria Decision-Making Approach and Deep Long Short-Term Memory Technique. *Sensors, 22*, 634.

Ali, M., Ono, N., Kaysar, M., Shamszaman, Z., Pham, T., Gao, F., . . . Mileo, A. (2017). Real-time data analytics and event detection for IoT-enabled communication systems. *Journal of Web Semantics, 42*, 19-37.

Aljawarneh, S. A., Vangipuram, R., Puligadda, V. K., & Vinjamuri, J. (2017). G-SPAMINE: An approach to discover temporal association patterns and trends in internet of things. *Future Generation Computer Systems, 74*, 430-443.

Al-Turjman, F., & Abujubbeh, M. (2019). IoT-enabled smart grid via SM: An overview. *Future Generation Computer Systems, 96*, 579-590.

Ambika, N. (2021). A Reliable Hybrid Blockchain-Based Authentication System for IoT Network. In S. Singh, & A. D. Jurcut(Eds.), *Revolutionary Applications of Blockchain-Enabled Privacy and Access Control* (pp. 219-233). USA: IGI Global.

Amri, M. F., Rizqyawan, M. I., & Turnip, A. (2016). ECG signal processing using offline-wavelet transform method based on ECG-IoT device. *3rd International Conference on Information Technology, Computer, and Electrical Engineering (ICITACEE)* (pp. 1-6). Semarang, Indonesia: IEEE.

Arridha, R., Sukaridhoto, S., Pramadihanto, D., & Funabiki, N. (2017). Classification extension based on IoT-big data analytic for smart environment monitoring and analytic in real-time system. *International Journal of Space-Based and Situated Computing, 7*(2), 82-93.

Awan, A. J., Brorsson, M., Vlassov, V., & Ayguade, E. (2015). Performance characterization of in-memory data analytics on a modern cloud server. *Fifth International Conference on Big Data and Cloud Computing* (pp. 1-8). Dalian, China: IEEE.

Baars, H., & Ereth, J. (2016). From Data Warehouses to Analytical Atoms–The Internet of Things as a Centrifugal Force in Business Intelligence and Analytics. *30 conference of european colloid and Interface society* (pp. 1-18). Rome: AIS Elibrary.

Babar, M., & Arif, F. (2019). Real-time data processing scheme using big data analytics in internet of things based smart transportation environment. *Journal of Ambient Intelligence and Humanized Computing, 10*(10), 4167-4177.

Babar, M., Alshehri, M. D., Tariq, M. U., Ullah, F., Khan, A., Uddin, M. I., & Almasoud, A. S. (2021). IoT-Enabled Big Data Analytics Architecture for Multimedia Data Communications. *Wireless Communications and Mobile Computing*, 1-9.

Bhatnagar, S., & Kumra, R. (2020). Understanding consumer motivation to share IoT products data. *Journal of Indian Business Research.*

Bhoi, S., Panda, S., Patra, B., Pradhan, B., Priyadarshinee, P., Tripathy, S., . . . Khilar, P. (2018). FallDS-IoT: a fall detection system for elderly healthcare based on IoT data analytics. *International*

Conference on Information Technology (ICIT) (pp. 155-160). Bhubaneswar, India: IEEE.

Bu, F., & Wang, X. (2019). A smart agriculture IoT system based on deep reinforcement learning. *Future Generation Computer Systems, 99*, 500-507.

Chang, K. M., Dzeng, R. J., & Wu, Y. J. (2018). An automated IoT visualization BIM platform for decision support in facilities management. *Applied sciences, 8*(7), 1086.

Chen, F., Deng, P., Wan, J., Zhang, D., Vasilakos, A. V., & Rong, X. (2015). Data mining for the internet of things: literature review and challenges. *International Journal of Distributed Sensor Networks, 11*(8), 431047.

Chen, W., Guo, F., & Wang, F. Y. (2015). A survey of traffic data visualization. *IEEE Transactions on Intelligent Transportation Systems, 16*(6), 2970-2984.

Chhaya, L., Sharma, P., Kumar, A., & Bhagwatikar, G. (2018). IoT-based implementation of field area network using smart grid communication infrastructure. *Smart Cities, 1*(1), 176-189.

Chin, W. L., Li, W., & Chen, H. H. (2017). Energy big data security threats in IoT-based smart grid communications. *IEEE Communications Magazine, 55*(10), 70-75.

Chowdury, M., Emran, T., Ghosh, S., Pathak, A., Alam, M., Absar, N., . . . Hossain, M. (2019). IoT based real-time river water quality monitoring system. *The 16th International Conference on Mobile Systems and Pervasive Computing (MobiSPC 2019), The 14th International Conference on Future Networks and Communications (FNC-2019);The 9th International Conference on Sustainable Energy Information Technology. 155*, pp. 161-168. Halifax, Canada.: ELSEVIER.

Chu, P. M., & Lee, S. J. (2017). A novel recommender system for E-commerce. *10th international congress on image and signal processing, biomedical engineering and informatics (CISP-BMEI)* (pp. 1-5). Shanghai, China: IEEE.

Coetzee, P., Leeke, M., & Jarvis, S. (2014). Towards unified secure on- and off-line analytics at scale. *Parallel Computing, 40*(10), 738-753.

Cui, Z., Jing, X., Zhao, P., Zhang, W., & Chen, J. (2021). A new subspace clustering strategy for AI-based data analysis in IoT system. *EEE Internet of Things Journal.*

Daneels, G., Municio, E., Spaey, K., Vandewiele, G., Dejonghe, A., Ongenae, F., . . . Famaey, J. (2017). Real-time data dissemination and analytics platform for challenging IoT environments. *Global Information Infrastructure and Networking Symposium (GIIS)* (pp. 23-30). Saint Pierre, France: IEEE.

Das, H., Dey, N., & Balas, V. E. (2019). *Real-time data analytics for large scale sensor data.* Cambridge, Massachusetts: Academic Press.

Derguech, W., Bruke, E., & Curry, E. (2014). An autonomic approach to real-time predictive analytics using open data and internet of things. *IEEE 11th Intl Conf on Ubiquitous Intelligence and Computing and 2014 IEEE 11th Intl Conf on Autonomic and Trusted Computing and 2014 IEEE 14th Intl Conf on Scalable Computing and Communications and Its Associated Workshops* (pp. 204-211). Bali, Indonesia: IEEE.

Dey, N., Hassanien, A. E., Bhatt, C., Ashour, A., & Satapathy, S. C. (2018). *Internet of things and big data analytics toward next-generation intelligence* (Vol. 35). Berlin: Springer.

Doan, Q. T., Kayes, A. S., R. W., & Nguyen, K. (2020). Integration of iot streaming data with efficient indexing and storage optimization. *IEEE Access, 8*, 47456-47467.

Du, J., Jiang, C., Gelenbe, E., Xu, L., Li, J., & Ren, Y. (2018). Distributed data privacy preservation in IoT applications. *IEEE Wireless Communications, 25*(6), 68-76.

Du, Z. (2020). Energy analysis of Internet of things data mining algorithm for smart green communication networks. *Computer Communications, 152*, 223-231.

Dwivedi, A. D., Srivastava, G., Dhar, S., & Singh, R. (2019). A decentralized privacy-preserving healthcare blockchain for IoT. *Sensors, 19*(2), 326.

Ediger, D., Jiang, K., Riedy, J., & Bader, D. A. (2010). Massive streaming data analytics: A case study with clustering coefficients. *International Symposium on Parallel & Distributed Processing, Workshops and Phd Forum (IPDPSW)* (pp. 1-8). Atlanta, GA, USA: IEEE.

Ekren, B. Y., Mangla, S. K., Turhanlar, E. E., Kazancoglu, Y., & Li, G. (2021). Lateral inventory share-based models for IoT-enabled E-commerce sustainable food supply networks. *Computers & Operations Research, 130*, 105237.

Elsaleh, T., Enshaeifar, S., Rezvani, R., Acton, S. T., Janeiko, V., & Bermudez-Edo, M. (2020). IoT-Stream: A lightweight ontology for internet of things data streams and its use with data analytics and event detection services. *Sensors, 20*(4), 953.

Faizan Ullah, M., Imtiaz, J., & Maqbool, K. Q. (2019). Enhanced three layer hybrid clustering mechanism for energy efficient routing in IoT. *Sensors, 19*(4), 829.

Farhan, M., Jabbar, S., Aslam, M., Ahmad, A., Iqbal, M. M., Khan, M., & Maria, M. E. (2018). A real-time data mining approach for interaction analytics assessment: IoT based student interaction framework. *International Journal of Parallel Programming, 46*(5), 886-903.

Fugini, M., Finocchi, J., & Locatelli, P. (2021). A Big Data Analytics Architecture for Smart Cities and Smart Companies. *Big Data Research, 24*, 100192.

García-Magariño, I., Nasralla, M. M., & Nazir, S. (2020). Real-time analysis of online sources for supporting business intelligence illustrated with bitcoin investments and iot smart-meter sensors in smart cities. *Electronics, 9*(7), 1-18.

Gaur, B., Shukla, V. K., & Verma, A. (2019). Strengthening people analytics through wearable IOT device for real-time data collection. *international conference on automation, computational and technology management (ICACTM)* (pp. 555-560). London, UK: IEEE.

Ge, Y., Liang, X., Zhou, Y. C., Pan, Z., Zhao, G. T., & Zheng, Y. L. (2016). Adaptive analytic service for real-time internet of things applications. *IEEE International Conference on Web Services (ICWS)* (pp. 484-491). San Francisco, CA, USA: IEEE.

Gluhak, A., Krco, S., Nati, M., Pfisterer, D., Mitton, N., & Razafindralambo, T. (2011). A survey on facilities for experimental internet of things research. *IEEE Communications Magazine, 49*(11), 58-67.

Goebel, M., & Gruenwald, L. (1999). A survey of data mining and knowledge discovery software tools. *ACM SIGKDD explorations newsletter, 1*(1), 20-33.

Gohar, M., Ahmed, S. H., Khan, M., Guizani, N., Ahmed, A., & Rahman, A. U. (2018). A big data analytics architecture for the internet of small things. *IEEE Communications Magazine, 56*(2), 128-133.

Gong, T., Huang, H., Li, P., Zhang, K., & Jiang, H. (2015). A medical healthcare system for privacy protection based on IoT. *Seventh International Symposium on Parallel Architectures, Algorithms and Programming (PAAP)* (pp. 217-222). Nanjing, China: IEEE.

Gupta, P., Agrawal, D., Chhabra, J., & Dhir, P. K. (2016). IoT based smart healthcare kit. *International Conference on Computational Techniques in Information and Communication Technologies (ICCTICT)* (pp. 237-242). New Delhi, India: IEEE.

Haseeb, K., Ud Din, I., Almogren, A., & Islam, N. (2020). An energy efficient and secure IoT-based WSN framework: An application to smart agriculture. *Sensors, 20*(7), 1-14.

Hassani, A., Medvedev, A., Zaslavsky, A., Delir Haghighi, P., Jayaraman, P., & Ling, S. (2019). Efficient Execution of Complex Context Queries to Enable Near Real-Time Smart IoT Applications. *Sensors, 19*, 5457.

He, J., Wei, J., Chen, K., Tang, Z., Zhou, Y., & Zhang, Y. (2017). Multitier fog computing with large-scale iot data analytics for smart cities. *IEEE Internet of Things Journal, 5*(2), 677-686.

Hopkins, J., & Hawking, P. (2018). Big Data Analytics and IoT in logistics: a case study. *The International Journal of Logistics Management., 29*(2).

Howard, A. J., Lee, T., Mahar, S., Intrevado, P., & Woodbridge, D. M. (2018). Distributed data analytics framework for smart transportation. *IEEE 20th International Conference on High Performance Computing and Communications; IEEE 16th International Conference on Smart City; IEEE 4th International Conference on Data Science and Systems (HPCC/SmartCity/DSS)* (pp. 1374-1380). Exeter, UK: IEEE.

Hu, H., & Tang, L. (2020). Edge Intelligence for Real-Time Data Analytics in an IoT-Based Smart Metering System. *IEEE Network, 34*(5), 68-74.

Hussain, M. M., Alam, M. S., & Beg, M. S. (2019). Fog computing model for evolving smart transportation applications. *Fog and Edge Computing: Principles and Paradigms, 22*(4), 347-372.

Islam, M. M., Rahaman, A., & Islam, M. R. (2020). Development of smart healthcare monitoring system in IoT environment. *SN computer science, 1*, 1-11.

Jan, B., Farman, H., Khan, M., Talha, M., & Din, I. U. (2019). Designing a smart transportation system: An internet of things and big data approach. *IEEE Wireless Communications, 26*(4), 73-79.

Jan, T., & Sajeev, A. S. (2018). Boosted probabilistic neural network for IoT data classification. *16th Intl Conf on Dependable, Autonomic and Secure Computing, 16th Intl Conf on Pervasive Intelligence and Computing,;4th Intl Conf on Big Data Intelligence and Computing and Cyber Science and Technology Congress*

(DASC/PiCom/DataCom/CyberSciTech) (pp. 408-411). Athens, Greece: IEEE.

Jeong, J. S., Han, O., & You, Y. Y. (2016). A design characteristics of smart healthcare system as the IoT application. *Indian Journal of Science and Technology, 9*(37), 52.

Jin, D. H., & Kim, H. J. (2018). Integrated understanding of big data, big data analysis, and business intelligence: a case study of logistics. *Sustainability, 10*(10), 3778.

Ju, Z., & Li, Y. (2011). Analysis on Internet of Things (IOT) based on the" Subway Supermarket" e-commerce mode of TESCO. *International Conference on Information Management, Innovation Management and Industrial Engineering* (pp. 430-433). Shenzhen, China: IEEE.

Khan, M., Silva, B. N., & Han, K. (2017). A web of things-based emerging sensor network architecture for smart control systems. *Sensors, 17*(2), 332.

Khoa, T. A., Man, M. M., Nguyen, T. Y., & Nguyen, V. N. (2019). Smart agriculture using IoT multi-sensors: A novel watering management system. *Journal of Sensor and Actuator Networks, 8*(3), 1-22.

Kim, J. Y., Lee, H. J., Son, J. Y., & Park, J. H. (2015). Smart home web of objects-based IoT management model and methods for home data mining. *17th Asia-Pacific Network Operations and Management Symposium (APNOMS)* (pp. 327-331). Busan, Korea (South): IEEE.

Kim, N. J., & Park, J. K. (2015). Sports analytics & risk monitoring based on hana platform: sports related big data & IoT trends by using HANA in-memory platform. *International SoC Design Conference (ISOCC)* (pp. 221-222). Gyeongju, Korea (South): IEEE.

Kodali, R. K., Swamy, G., & Lakshmi, B. (2015). An implementation of IoT for healthcare. *Recent Advances in Intelligent Computational Systems (RAICS)* (pp. 411-416). Trivandrum, India: IEEE.

Krishnamurthi, R., Kumar, A., Gopinathan, D., Nayyar, A., & Qureshi, B. (2020). An Overview of IoT Sensor Data Processing, Fusion, and Analysis Techniques. *Sensors, 20*, 6076.

Lakshmanaprabu, S. K., Shankar, K., Ilayaraja, M., Nasir, A. W., Vijayakumar, V., & Chilamkurti, N. (2019). Random forest for big data classification in the internet of things using optimal features. *International journal of machine learning and cybernetics, 10*(10), 2609-2618.

Li, Z., Liu, G., Liu, L., Lai, X., & Xu, G. (2017). IoT-based tracking and tracing platform for prepackaged food supply chain. *Industrial Management & Data Systems., 117*(9).

Liou, D. Y., Chen, C. C., Chen, T. F., & Lin, T. J. (2017). Accelerating R data analytics in IoT edge systems by memory optimization. . *11th International Conference on Application of Information and Communication Technologies (AICT)* (pp. 1-5). Moscow, Russia: IEEE.

Liu, X., & Nielsen, P. S. (2015). Streamlining smart meter data analytics. *10th Conference on Sustainable Development of Energy, Water and Environment Systems. 558*, pp. 1-14. Dubrovnik, Croatia: University of Zagreb;Instituto Superior Técnico.

Luque-Vega, L. F., Michel-Torres, D. A., Lopez-Neri, E., Carlos-Mancilla, M. A., & González-Jiménez, L. E. (2020). Iot smart parking system based on the visual-aided smart vehicle presence sensor: SPIN-V. *Sensors, 20*(5), 1476.

Lv, Z., & Qiao, L. (2020). Analysis of healthcare big data. *Future Generation Computer Systems, 109*, 103-110.

Lv, Z., Lou, R., Li, J., Singh, A. K., & Song, H. (2021). Big data analytics for 6G-enabled massive internet of things. *IEEE Internet of Things Journal, 8*(7), 5350-5359.

Malekzadeh, M., Clegg, R. G., & Haddadi, H. (2018). Replacement autoencoder: A privacy-preserving algorithm for sensory data analysis. *IEEE/acm third international conference on internet-of-things design and implementation (iotdi)* (pp. 165-176). Orlando, FL, USA: IEEE.

Marjani, M., Nasaruddin, F., Gani, A., Karim, A., Hashem, I. A., Siddiqa, A., & Yaqoob, I. (2017). Big IoT data analytics: architecture, opportunities, and open research challenges. *ieee access, 5*, 5247-5261.

Mekala, M. S., & Viswanathan, P. (2017). A novel technology for smart agriculture based on IoT with cloud computing. *International Conference on I-SMAC (IoT in Social, Mobile, Analytics and Cloud)(I-SMAC)* (pp. 75-82). Palladam, India: IEEE.

Muangprathub, J., Boonnam, N., Kajornkasirat, S., Lekbangpong, N., Wanichsombat, A., & Nillaor, P. (2019). IoT and agriculture data analysis for smart farm. Computers and electronics in agriculture. *Computers and electronics in agriculture, 156*, 467-474.

Muthu, B., Sivaparthipan, C. B., Manogaran, G., Sundarasekar, R., Kadry, S., Shanthini, A., & Dasel, A. (2020). IOT based wearable sensor for diseases prediction and symptom analysis in healthcare sector. *Peer-to-peer networking and applications, 13*(6), 2123-2134.

Muthuramalingam, S., Bharathi, A., Gayathri, N., Sathiyaraj, R., & Balamurugan, B. (2019). IoT based intelligent transportation system (IoT-ITS) for global perspective: A case study. In B. V., S. V., K. R., & K. M. (eds), *Internet of Things and Big Data Analytics for Smart Generation* (pp. 279-300). Cham: Springer.

Nagaraj, A. (2021). *Introduction to Sensors in IoT and Cloud Computing Applications.* UAE: Bentham Science Publishers.

Ozawa, S., Ban, T., Hashimoto, N., Nakazato, J., & Shimamura, J. (2020). A study of IoT malware activities using association rule learning for darknet sensor data. *International Journal of Information Security, 19*(1), 83-92.

Pal, A., & Kant, K. (2018). IoT-based sensing and communications infrastructure for the fresh food supply chain. *computer, 51*(2), 76-80.

Parvin, S., Venkatraman, S., Souza-Daw, T. d., Fahd, K., Jackson, J., Kaspi, S., . . . Gawanmeh., A. (2019). Smart food security system using iot and big data analytics. *16th International Conference on Information Technology-New Generations (ITNG 2019)* (pp. 253-258). Las Vegas, Nevada, USA: Springer, Cham.

Pastor-Vargas, R., Tobarra, L., & Robles-Gómez, A. M. (2020). A WoT Platform for Supporting Full-Cycle IoT Solutions from Edge to Cloud Infrastructures: A Practical Case. *Sensors, 20*(13), 3770.

Patel, P., Ali, M. I., & Sheth, A. (2017). On using the intelligent edge for IoT analytics. *IEEE Intelligent Systems, 32*(5), 64-69.

Patil, K. A., & Kale, N. R. (2016). A model for smart agriculture using IoT. *2016 International Conference on Global Trends in Signal Processing, Information Computing and Communication (ICGTSPICC)* (pp. 543-545). Jalgaon: IEEE.

Pawar, P. (2019). Design and development of advanced smart energy management system integrated with IoT framework in smart grid environment. *Journal of Energy Storage, 25,* 100846.

Puschmann, D., Barnaghi, P., & Tafazolli, R. (2016). Adaptive clustering for dynamic IoT data streams. *IEEE Internet of Things Journal, 4*(1), 64-74.

Quan, X. X., Yang, J. F., & Luo, Z. (2021). Models in digital business and economic forecasting based on big data IoT data visualization technology. *Personal and Ubiquitous Computing*, 1-7.

Rahmani, A. M., Babaei, Z., & Souri, A. (2021). Event-driven IoT ar-chitecture for data analysis of reliable healthcare application using complex event processing. *Cluster Computing, 24*(2), 1347-1360.

Rathore, M. M., Ahmad, A., Paul, A., & Jeon, G. (2015). Efficient graph-oriented smart transportation using internet of things gener-ated big data. *11th International Conference on Signal-Image Technology & Internet-Based Systems (SITIS)* (pp. 512-519). Bangkok, Thailand: IEEE.

Rathore, M. M., Ahmad, A., Paul, A., & Thikshaja, U. K. (2016). Ex-ploiting real-time big data to empower smart transportation using big graphs. *IEEE Region 10 Symposium (TENSYMP)* (pp. 135-139). Bali, Indonesia: IEEE.

Rathore, M. M., Paul, A., Hong, W. H., Seo, H., Awan, I., & Saeed, S. (2018). Exploiting IoT and big data analytics: Defining smart digital city using real-time urban data. *Sustainable cities and society, 40*, 600-610.

Ray, P. P., Dash, D., & De, D. (2020). Real-time event-driven sensor data analytics at the edge-Internet of Things for smart per-sonal healthcare. *The Journal of Supercomputing, 76*(9), 6648-6668.

Reda, R., Piccinini, F., & Carbonaro, A. (2018). Towards consistent data representation in the IoT healthcare landscape. *Proceed-ings of the 2018 International Conference on Digital Health* (pp. 5-10). ACM: Lyon France.

Renart, E. G., Veith, A. D., Balouek-Thomert, D., De Assunção, M. D., Lefevre, L., & Parashar, M. (2019). Distributed operator placement for IoT data analytics across edge and cloud re-sources. *19th IEEE/ACM International Symposium on Cluster, Cloud and Grid Computing (CCGRID)* (pp. 459-468). Larnaca, Cyprus: IEEE.

Rghioui, A., Lloret, J., & Oumnad, A. (2020). Big data classification and internet of things in healthcare. *International Journal of E-Health and Medical Communications (IJEHMC), 11*(2), 20-37.

Ruan, J., & Shi, Y. (2016). Monitoring and assessing fruit freshness in IOT-based e-commerce delivery using scenario analysis and interval number approaches. *Information Sciences, 373*, 557-570.

Ryu, M., Kim, J., & Yun, J. (2015). Integrated semantics service plat-form for the Internet of Things: A case study of a smart office. *Sensors, 15*(1), 2137-2160.

Saikiran, P., SureshBabu, E., Padmini, D., SriLalitha, V., & Krishnnand, V. (2017). Security issues and countermeasures of three tier architecture of IoT-a survey. *International Journal of Pure and Applied Mathematics, 115*(6), 49-57.

Sakr, S., & Elgammal, A. (2016). Towards a comprehensive data analytics framework for smart healthcare services. *Big Data Research, 4*, 44-58.

Salahuddin, M. A., A. Al-Fuqaha, M. G., Shuaib, K., & Sallabi, F. (2017). Softwarization of Internet of Things Infrastructure for Secure and Smart Healthcare. *Computer, 50*(7), 74-79.

Saleem, T. J., & Chishti, M. A. (2019). Deep learning for Internet of Things data analytics. *16th Learning and Technology Conference 2019Artificial Intelligence and Machine Learning: Embedding the Intelligence. 163*, pp. 381-390. Jeddah, Saudi Arabia.: ELSEVIER.

Sanjeevi, P., Prasanna, S., Siva Kumar, B., Gunasekaran, G., Alagiri, I., & Vijay Anand, R. (2020). Precision agriculture and farming using Internet of Things based on wireless sensor network. *Transactions on Emerging Telecommunications Technologies, 31*(12), p.e3978.

Santos, M., e Sá, J., Costa, C., Galvão, J., Andrade, C., Martinho, B., . . . Costa, E. (2017). A big data analytics architecture for industry 4.0. *World Conference on Information Systems and Technologies* (pp. 175-184). Madeira, Portugal: Springer, Cham.

Saranya, S. S., & Fatima, N. S. (2020). Context Aware Data Fusion on Massive IOT Data in Dynamic IOT Analytics. *Webology, 17*(2), 957-970.

Savaglio, C., Gerace, P., Di Fatta, G., & Fortino, G. (2019). Data mining at the IoT edge. *28th International Conference on Computer Communication and Networks (ICCCN)* (pp. 1-6). Valencia, Spain: IEEE.

Shadroo, S., & Rahmani, A. M. (2018). Systematic survey of big data and data mining in internet of things. *Computer Networks, 139*, 19-47.

Shao, C., Yang, Y., Juneja, S., & GSeetharam, T. (2022). IoT data visualization for business intelligence in corporate finance. *Information Processing & Management, 59*(1), 102736.

Singh, S., & Singh, N. (2015). Internet of Things (IoT): Security challenges, business opportunities & reference architecture for E-commerce. *International Conference on Green Computing and*

Internet of Things (ICGCIoT) (pp. 1577-1581). Greater Noida, India: IEEE.

Sushanth, G., & Sujatha, S. (2018). IOT based smart agriculture system. *International Conference on Wireless Communications, Signal Processing and Networking (WiSPNET)* (pp. 1-4). Chennai, India: IEEE.

Ta-Shma, P., Akbar, A., Gerson-Golan, G., Hadash, G., Carrez, F., & Moessner, K. (2017). An ingestion and analytics architecture for iot applied to smart city use cases. *IEEE Internet of Things Journal, 5*(2), 765-774.

Tavera Romero, C. A., Ortiz, J. H., Khalaf, O. I., & Ríos Prado, A. (2021). Business intelligence: business evolution after industry 4.0. *Sustainability, 13*(18), 10026.

Tawalbeh, L. A., Muheidat, F., Tawalbeh, M., & Quwaider, M. (2020). IoT Privacy and security: Challenges and solutions. *Applied Sciences, 10*(12), 4102.

Tönjes, R., Barnaghi, P., Ali, M., Mileo, A., Hauswirth, M., Ganz, F., . . . Puiu, D. (2014). Real time iot stream processing and large-scale data analytics for smart city applications. *European Conference on Networks and Communications* (p. 10). Bologna, Italy: IEEE Xplore.

Tsai, C. W., Lai, C. F., Chiang, M. C., & Yang, L. T. (2013). Data mining for internet of things: A survey. *IEEE Communications Surveys & Tutorials, 16*(1), 77-97.

Vass, T. d., Shee, H., & Miah, S. J. (2020). Iot in supply chain management: a narrative on retail sector. *INTERNATIONAL JOURNAL OF LOGISTICS: RESEARCH AND APPLICATIONS*, 1-20.

Vázquez-Gallego, F., Tuset-Peiró, P., Alonso, L., & Alonso-Zarate, J. (2020). Delay and energy consumption analysis of frame slotted ALOHA variants for massive data collection in internet-of-things scenarios. *Applied Sciences, 10*(1), 327.

Verma, S., Kawamoto, Y., Fadlullah, Z. M., Nishiyama, H., & Kato, N. (2017). A survey on network methodologies for real-time analytics of massive IoT data and open research issues. *IEEE Communications Surveys & Tutorials, 19*(3), 1457-1477.

Wu, C. Y., & Huang, K. H. (2020). A Framework for Off-Line Operation of Smart and Traditional Devices of IoT Services. *Sensors, 20*(21), 6012.

Wu, X. Q., Zhang, L., Tian, S. L., & Wu, L. (2021). Scenario based e-commerce recommendation algorithm based on customer interest in Internet of things environment. *Electronic Commerce Research, 21*(3), 689-705.

Xu, C., Ren, J., Zhang, D., & Zhang, Y. (2018). Distilling at the edge: A local differential privacy obfuscation framework for IoT data analytics. *IEEE Communications Magazine, 56*(8), 20-25.

Yacchirema, D. C., Sarabia-Jácome, D., Palau, C. E., & Esteve, M. (2018). A smart system for sleep monitoring by integrating IoT with big data analytics. *IEEE Access, 6*, 35988-36001.

Yan, J., Liu, J., & Tseng, F. M. (2020). An evaluation system based on the self-organizing system framework of smart cities: A case study of smart transportation systems in China. *Technological Forecasting and Social Change, 153*, 1-12.

Yao, X., Wang, J., Shen, M., Kong, H., & Ning, H. (2019). An improved clustering algorithm and its application in IoT data analysis. *Computer Networks, 159*, 63-72.

Zeadally, S., Siddiqui, F., Baig, Z., & Ibrahim, A. (2019). Smart healthcare: Challenges and potential solutions using internet of things (IoT) and big data analytics. *PSU research review, 4*(2), 93-109.

Zgank, A. (2020). Bee swarm activity acoustic classification for an IoT-based farm service. *Sensors, 20*(1), 21.

Zhang, J., Wang, Y., Li, S., & Shi, S. (2020). An Architecture for IoT-Enabled Smart Transportation Security System: A Geospatial Approach. *IEEE Internet of Things Journal, 8*(8), 6205-6213.

Zhang, L., Vinodhini, B., & Maragatham, T. (2021). Interactive IoT Data Visualization for Decision Making in Business Intelligence. *Arabian Journal for Science and Engineering*, 1-11.

Zhang, Y., Ren, S., Liu, Y., & Si, S. (2017). A big data analytics architecture for cleaner manufacturing and maintenance processes of complex products. *Journal of cleaner production, 142*, 626-641.

Zheng, S., Apthorpe, N., Chetty, M., & Feamster, N. (2018). User perceptions of smart home IoT privacy. *Proceedings of the ACM on Human-Computer Interaction. 2*, pp. 1-20. New York, NY, United States: ACM.

Zhu, L. (2020). Optimization and simulation for e-commerce supply chain in the internet of things environment. *Complexity*.

3

SENSORS AND DATA ANALYTICS

Abstract

Sensors are tiny elements deployed in the environment to monitor any object of interest. These devices sense the atmosphere and collect the facts. They process them and dispatch them to the pre-defined location. An enormous amount of data is examined using different algorithms. The chapter details the sensors, architecture, applications, challenges faced by the technology, and future directions.

3.1 Introduction to sensor data analytics

The detectors (Ambika N. , 2020) (Yick, Mukherjee, & Ghosal, 2008) are small devices embedded in the environment to sense the surroundings and provide the readings to the pre-defined destination. They have a sourcing element and a receiver. These devices employ distributed computing. They form the network dynamically. The association between machines relies on their place and area deviation over the period. They deliver assistance in the transmission of data.

The information investigation (Hariri, Fredericks, & Bowers, 2019) explains the procedure of examining enormous databases. It uncovers practices, unfamiliar associations, demand tendencies, client priorities, and other beneficial details. The sensors collect information. It tracks the object of interest and provides reading to the destination. The facts supply actionable knowledge and enhance conclusion-making capacities for institutions and organizations.

3.2 Sensor architecture

The user spreads the detectors in the environment. Every device qualifies to accumulate information and course back knowledge to the base station. Facts move to the destination using multi-hop methodology. The source can intercommunicate with the assignment supervisor using cyberspace or Satellites. To make the system better various architecture (Gajbhiye & Mahajan, 2008) detailing is necessary. The strategy focuses on diverse characteristics like imperfection forbearance, flexibility, dimension, rich set of functions, manufacturing prices, etc. This section details the same.

Wireless detector systems (Ambika, 2020) (Raghavendra, S., Sivalingam, & Znati, 2006.) need new possibilities for omnipresent surveillance of the surroundings. The method allows inventors to monitor, investigate, and question disseminated detector information at numerous decisiveness. It controls spatial and temporal association. Many queries over these systems will be spatial chronological. The warehouse should help competent probing and excavating possibilities of curiosity. Clients will habitually need compact outlines of extensive knowledge. These requirements lead to the usage of different architectures.

The system (Dias, Adame, Bellalta, & Oechsner, 2016) has detector hardware constraints and a detachment from the information source to the prominent host. The framework details sensed information. The data undergoes grouping, narration, creating warehouse, supervision, exploration, imaging, and investigation. The suggested approach has fact examination for the sensing console. It has intersected elements that can swap details with authorized objects of various disciplines. The transmission between the dashboard and exterior segment tracks a traditional approach. The system obtains outer information from web-based demands. It reports respective information in the state of occurrences. The control panel creates broadcast information using an isolated entry system. In the study, the sensor network has a gateway and four detectors. They operate unassisted of each other. The detector devices survey interior and outside surroundings. It installs in a structured apparatus without fuel with a dedicated cyber relationship and explicit admission to the framework elements. A cyber client podium delivers the gathered knowledge and notifies periodic defeats. It also results from a detailed investigation conducted by an exterior hub. The information examination gateway is in the R program vocabulary. The analysis uses details pertinence estimation and prediction.

The recommended framework (Aydin, Hallac, & Karakus, 2015) has three segments. The study senses knowledge using the detectors in the data harvesting stage. The study collects data in the data storage phase. The work performs investigation in the data analysis phase. The application medium is Sun Fire X4450 hubs with 24 calibration centers of Intel 3.16 GHz CPU and 64 GB of recall cells. It employs Ubuntu 14.04 as the organizing design. GPS earpieces ascended on the automobiles can register their area via GPRS. The detectors open a link to the TCP gateway in many conditions. The work uses Quick-Server, an open-source Java bookstore. It provides immediate outcomes and supports multiclient TCP hub presentations. It has a secluded management podium called QSAdminServer. It addresses all the doings of the gateway programs. Information accumulated from the sensing elements is in the repository. The method uses MongoDB to store data. It computes the details using Openstack. The system uses disseminated computer-learning procedures. Apache Mahout and MLLib by Apache Spark are open sources dispersed architecture for information examination. The work uses the K-means approach for grouping two-dimensional GPS position learning.

The MULE framework (Shah, Roy, Jain, & Brunette, 2003) delivers wide-area connections for a dispersed detector system. It explores the usage of active entities like populace and automobiles existing in the working surrounding. These portable devices serve as an in-between tier. It comprises a package carrier method to link the system. It makes a three-phase generalization level. It regulates various circumstances. The first layer has entry points that transfer the information stored on a MULE to the WAN. The devices are at suitable places where grid connection and fuel are unrestricted. Entrance ports can transmit with a central information repository allowing them to coordinate the facts that they accumulate, notice replications, and yield responses. The center level has transportable vendors that supply the web with links. The procedure is stretchable for additional resources and elastic with a low price. The bottommost deck of the system has unsystematically disseminated detectors. The devices intercommunicate employing a short-range receiver. The energy and recall cells are some restrictions.

The foremost step (Borgman, Wallis, Mayernik, & Pepe, 2007) recycles information from previous investigations to develop new experimentations. Curious areas or time duration for knowledge clusters are recognized. The phase comprises setting the detectors to install. They regulate with available answers. The work controls detectors in the workshop on the marine living hoped to be in the water. The influence of climate is handled by sensing components in regional necessities. The boundary parameters are analyzed to manage other possible detector relics. The crew starts to accumulate clarification of biological occurrences from the detecting elements. The exploring group tests the compliances after being seized to review for knowledge trustfulness, detector dependability, inconsistencies, and other characteristics. They can notice the dangerous compounds. These sorts of keeping functions are intake to examples. Center for Embedded Networked Sensing appliances are devising forms to enhance knowledge by recognizing and fixing mistakes. Detector facts combine with hand-accumulated knowledge. Water specimens are four times during 24 hours, whereas water detectors may grasp 4 data points per minute. Learning examination happens after the information is confirmed, obtained, incorporated, and cleansed. Research units use numerical demonstration and presentation instruments. The sample and develop assumptions and make decisions about knowledge acquired from the installation. Details from implanted detector terminate in writings. Some information is in a new Center for Embedded Networked Sensing warehouse, known as SensorBase.

The framework has base station communication of detector information in clothing devices. It needs ongoing real-time revises of up to 100 Hz. The design (Benbasat, Morris, & Paradiso, 2003) has panels of 1.4 inches and 0.4 inches in height. They connect using electricity by two titles with 26 pins at contrasting intersections. The connectives are Molex Milli-Grid wrapped headings and mating plates. The other two junctions climb spots that qualify for structural mounting of the pile. The mainboard has 22 MIPS processing components with 12-bit ADC and 115.2 kBps 916 MHz sender. This delivers for a disseminated multiplexer bus with location feed. It allows connections. It is the process for detailed transmission from the detector panels. Information communicates using the link. It is a multipurpose input/output or exterior intersect pins on the processing element. It is accountable for the assemblage and communication to the prominent hub. Energy ordinance controls outwardly. The battery used is +5V and +3V. The ground ties to the primary panel. A gateway uses a primary podium. It qualifies as a TDMA wireless procedure. Acceleration organizes using two Analog instruments. It is orthogonally connected to the side of the pane to reach the third axis of perception. The momentum is estimated using two Murata ENC03J gyroscopes and a single Analog Devices ADXRS150 gyroscope. This blend permits for 6-axis inertial detection. The receiver panel transmits a single 40 kHz vibration. A third detector forum permits various tactile and stress detectors. It contains four single-ended force-sensitive resistors, two differential bend detectors, and two piezoelectric sensing elements. It includes the circuitry for a 9-channel loading method capacitive detector.

The architecture (Stripelis, Ambite, Chiang, Eckel, & Habre, 2017) uses Apache Kafka and Apache Spark. It incorporates detector and conventional information repositories. The mediation tier operates on mappings and questioning updating. It connects the plan from the information repositories (varied) into a blended design for the investigating associates. The *knowledge warehouse* slab has existing sensing elements and cyber benefits that feed information to the procedure. It has obtained learning from diverse origins. The *Streaming* tier gets details from executing detectors utilizing Apache Kafka. The element connects to a single Kafka fabricator. It registers the input details to a specified Kafka subject. The *Integration* slap associates the knowledge from the diverse origins into a single design employing analytical association and question update. The Apache Spark SQL motor regulates the dispersal of the questions and forces unprocessed sensing information and coordinated details to the Depository tier. The *Storage* slab achieves an enduring learning warehouse. The PRISMS-DSCIC creates a database for prospective investigations of pediatric breathing problems. The system stores data in HDFS and PostgreSQL. The *Analytics* tier constructs SparkR and Spark MLlib bookstore to allow investigators to examine real-time and recorded information.

The framework (Fidaleo, Nguyen, & Trivedi, 2004) has six elements. The host layer gives authority to random customers in an extensible structure. The organizer allows mining, broadcasting, and computation of information from numerous origins of knowledge. These knowledge repositories possess information obtained from climate, acoustic transducers, GPS, RFID, and seismic detectors. Confidential cleaning and Safety are critical. The frame transmission tracks are protected to control admission to mediator and outcomes of an investigation by illegal clients. Detectors and Hardware medium is where sensing elements attach to the structure. The understanding component has documented knowledge and masterful data. Workout presentation is a collaborating 3D podium for command and straining mixed detecting elements. The software medium permits a worker to define dimensional regions of curiosity. Simulated sensing elements are non-specified programs working on genuine details. These consumers employ occurrence inferencing, knowledge compaction, and association.

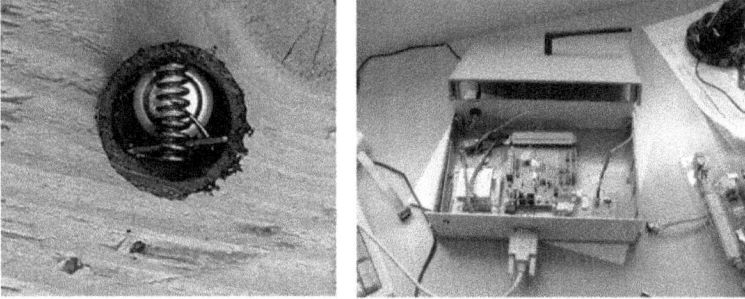

Figure 3.1 (a) Node inserted in the wood; figure 3.1 (b) Internal arrangement of sink components (Capella, Perles, Bonastre, & Serrano, 2011).

The work (Capella, Perles, Bonastre, & Serrano, 2011) considers measuring Environment surveillance. It gauges the wetness and warmth of the surroundings. It calculates the balance dampness of the timber. The readings notice reflection variations on the detection of Flies. The component uses LEDs and sunlight detectors. The embedded execution is star composition. The end devices are 2.5 × 5 cm implanted in lumber. The hole closed employing a timber lid to keep interior situations. The instruments gather data and transmit it to the server. The system uses a Silabs C8051F930 low-power microcontroller. It calculates the equalizer wetness of the lumber. The setup has an attractor for creatures. The component detects them employing sunlight reflecting deviations made with a LED from Avago and a perceptiveness detector. The system energies the devices using high fuel viscosity 1,100 mAh, 3.6 V. lithium-thionyl battery. The RF division operates a Texas Instruments/Chipcon CC1101 ISM band transceiver. The chip organizes for the 868 MHz European ISM band. The architecture assembles details and transmits them utilizing a GSM/GPRS Telit GM-862 to secluded surroundings. Figure 3.1 portrays the same.

3.3 Applications

3.3.1 Healthcare

The scientific instruments employ the current vesture as they give uneasiness and restrict observation of living in the workshop. The wearables (Lai, Palaniswami, & Begg, 2011) (Yang, 2006) substitute the unmanageable, inflexible plastics and alloy elements in the detectors. The microelectronics with membrane-like fabrics improve usage and functionality. Clothing supple mixed microchip technology has attributes that coordinate to the mortal. It symbolizes a pattern modification in habiliment. It ranges from inaccurately to carefully associated podium for boosted fitness surveillance, diagnostics, and human-motor medium. Figure 3.2 portrays the taxonomy of wearable sensor technology.

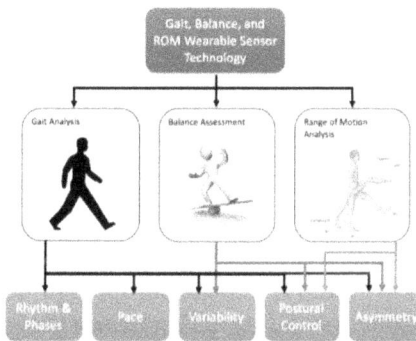

Figure 3.2 Taxonomy of Wearable Sensor Technology for Gait, Balance, and ROM Research (Díaz, Stephenson, & Labrador, 2020).

The sensing elements (Ge, et al., 2020) gather the patient's information. The subordinate devices node encodes the sick knowledge and upload it into the stockpile. The suffering information undergoes translation using deep knowledge procedures. The sick has to satisfy the need to enter the stored learning. The framework decodes the ciphertext of the inserted private credential. The suffering can erase the information on the repository by sending a request to the administration component. The system acknowledges the patient for the request received. The warehouse yields a confirmation notification after accepting an information return demand. The suffering then transmits a re-encoding credential to the stock. It produces a new source of the MHT to the sick after computing the omission. The long-suffering employs the information from the stockpile and the origin to confirm that the erasure was thriving. The details from the store-house operate on storage, designating education prototypes, fore-casting the potential infections of sicks, and cautioning the probable disorders. The ciphertext kept is transmitted to a high-performance host collection whose attributes meet the decoding standards. The details translate to a high-performance gateway grouping and exer-cise. The high-performance governed by the infirmary and approved by the physician will copy and interpret the knowledge from the stor-age and utilize the deciphered information to teach the prototype and create foretelling. The anticipated outcome ships to the medic conse-quent to the suffering, and the medic directions or alerts the sick based on the expected outcome. The fitness unit confines the entry system. The credential allocated to the physician possesses the re-lated admission framework. Competent medics interpret the same.

The mPHASiS (Kulkarni & Ozturk, 2011) host employs a specific in-vestigation. The instrument supervisor keeps machine administrator authorization. It copies the detector information to the gateway and duplicates sensing information composition from the hub. The man-ager can add, erase or watch a customer, simulate or stop assistance. It supplies the information to trigger benefits. The doctor can insert occurrences, organize and erase occasions, logging/log out of the pro-cedure, and portray the details to the sick. The Nursemaid can view the history of the respective patients. The framework has two play-ers. The portable customers get attention from the system. mPHASiS hub copy the checklist of sick with or without watches. The mPHASiS gateway framework is a traditional three-tier. It has a visualization phase, middle-layer, and dataset level. A multi-layer frame permits the detachment of circumstances where any class in the procedure can be extended and revised with low effect. The deck promotes high consistency and low connection between the levels. The middle layer splits into the industry and net levels. The machine administrator and portable customer convey with mPHASiS hub enterprise level over a TCP/IP and HTTP association. The net level connects with the presen-tation deck.

The mHealth application (Lee, Walker, Kim, & Lee, 2021) utilizes three players. Two parties were nutritious. The third player has a backache. The individual did not have any issues with strolling. The contestants sported the mobile. The device was in a pouch of a belt in the middle of their bodies. The applicants were advised to stride at their speed in an interior room of a structure. They stepped a length of about 400 feet. They hiked for about 2-3 minutes and then resumed to the initial position. The time of the experiment was 4-5 minutes. They were permitted to take the test once. The device kept the movement detector information in the application. It gathered the sensing knowledge in 20 milliseconds. The system kept the perceived details in an SQLite repository. The work instructed the participant to use the start and stop buttons. The experiment has a rotation matrix utilized to define the strolling outcome.

Fitness assistance application (Cortés, Bonnaire, Marin, & Sens, 2015) is on mobile devices. It is with detectors. The work concentrates on the services given by the system. Endomondo is a famous athletic-oriented program with 30 million clients. It authorizes clients to follow and convey their activity outcomes with mates. It broadcasts the same to the public domain. The shared exercises are a rich origin of real-world hints of athletic movements. The software operates on portable devices, employs detectors such as GPS and accelerometers to follow the machine along its course, and reports cumulative conditions. It can be improved with customer remarks and conveyed through the network. The database has the game actions diary of 15090 clients selected arbitrarily. It has 333689 pieces of training received over five months in 2014. Every catalog possesses client data, a training outline, and a GPS trail. The customer dataset holds details of the consumers. The activity overview shows statistics evaluated during the treatment. The GPS imprint has a GPS track developed along the pathway of the exercise duration.

The system (Alsiddiky, Awwad, Fouad, Hassanein, & Soliman, 2020) has three phases of bottleneck surveillance and supervision. The blockage administration depends on the governance procedure for facilitating communications and warehouse use. The suggested prototype delivers the comparative preference in diverse bottleneck phases. The component dispatches the communication details. It receives on a dynamic column administration procedure. It determines the situation of information at the introductory stage. It collaborates the assortment relevant priority to preserve the traffic measures. It enhances the information communication methodology. It uses representative information administration for controlling the communication rate. The knowledge is acquired haphazardly and shipped in the communication route. It organizes the forthcoming information in the column based on preference. The fuzzy logic system integrates with altered traffic metrics. The traffic category expands to support the highest quantity of information. The streaming data recovery uses a traversing tree to decrease the wavering during information communication. It notices the existence and scarcities of the five categories. The proposed system creates a traversing tree and organizes the time duration. The machine reports the detector to transmit the facts. It regulates the medium entrance. The last phase is decision-making. This work suggests the assumption design in the machine to estimate the communication of the details. It delivers dependent decision-making in which the devices can order the information from doctors.

The derivative (Yan, Huo, Xu, & Gidlund, 2010) is incorporated in adult's household. It contains grids incorporated machines. It has two phases. The foremost step has an unconnected detector system gathering the surrounding and confidential attributes. The next phase has information transmission to the system located at various distance. The host is the entrance gateway to the detached framework. The sensor panel embedding multiple pins has a segment aggregating and communicating the atmosphere sensed information. The component can serve as a router redirecting gathered details. The panel connects to various sensing elements. It gets beacons from these devices. The panel provisions force, heart pace, blaze warning, and crisis controller assemblage. The conflagration devices are in the pantry. The detector can notice the blaze sends the alert to the gateway. It generates a signal to warn the populace in the area. The vibration machine estimates the mortal heart pace.

Omnipresent fitness surveillance (Dey, Ashour, Shi, Fong, & Sherratt, 2017) helps maintain Patient confidential health conditions and doctor therapy reports. It delivers a first help reply in heart detention issues. A mobile machine connects to the individual having unconnected-based computation. It ships the perceived ECG sign. The ECG beacon, heart rate, and vibration speed of the patient. The measure gives a notification to the doctor or the nearby clinic in case of any discrepancies. Mobile transmission or cyberspace associates the sick in the heart detention issue. The Zigbee Alliance designed the Zigbee composition for many technologies. It maintains low-cost association and low fuel for gadgets that work on batteries. The attributes have IEEE 802.15.4 benchmarks.

The Body sensor network (Lo, Thiemjarus, King, & Yang, 2005)employ Texas Instrument MSP430 16-bit ultra-low power RISC calibration unit with 60KB+256B Flash recall cells, 2KB RAM, 12-bit ADC, and six analog tracks. The unconnected prototype has an outcome of 250kbps with a capacity of over 50m. 512KB serial flash memory is in the sensor of the information warehouse. The device executes on TinyOS by U.C. Berkeley. It has a tiny, publicly available, and fuel-efficient panel. It supplies a collection of structured programs. The developers could select the elements they need. The dimensions of resources are 200 bytes. The suggestion is the current strategy of the sensor instruments and their association with another sensor podium. It delivers a hands-on understanding of employing the devices for making an unconnected setup.

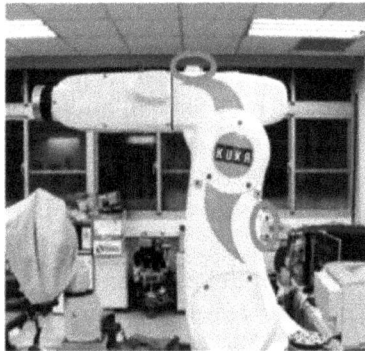

Figure 3.3 The robotic arm for sensor calibration. (Chiang, Chen, Liu, Hsu, & Chan, 2017).

The work (Chiang, Chen, Liu, Hsu, & Chan, 2017) considers seed for the assessment of knee ROM. It investigates the association between the knee ROM and the sick's situation. It uses a detector expanse from the knee link and the tilt of the physique part to estimate the movement curves of the knee intersection. The device positions itself on the thigh and the leg. The work positions them on the frame piece. It documents the movement of the thigh or the shank. The outcome splits into translational and gravitational segments. The gestures from accelerometers compute the curves of hip flexion and knee flexion. The work takes care of zero glide problems of the detectors and the precision guarantee. The tiny instruments can robotically execute self-alteration used in programs designed for client mediators. The inner computation emphasizes the mechanical associate as a traditional means to cross-verify the machine's exactness. Figure 3.3 represents the same.

3.3.2 Precision farming

Agriculturalists employ intelligent machines to switch on and off sprayers, aid in surveillance to estimate yield amount, ground wetness detectors, earth viscosity devices, infrared produce fitness sensing machines. This identifying approach cleverly constructs farmsteads. They are known as intelligent husbandry (Venkatramanan, Shah, & Prasad, 2020). Accurateness farming is an agribusiness idea having perceiving, calculating, and reacting to outer/inside domain's inconsistency of produce. The objective of accuracy husbandry investigation describes a finding aid technique for the system to balance recoveries on intakes while maintaining necessities. The system contains detectors for ground wetness, atmosphere temperature, moisture and stress, drizzle, and hail. Surveillance report, delivered by the sensing elements, is forced to a consolidated commodity. The information is enhanced and investigated. Outcomes of the research are transmitted to farmstead workers.

The intelligent husbandry model (Suakanto, Engel, Hutagalung, & Angela, 2016) has grid detectors and gateways. The devices accumulate information. Grid machines are a cluster of tiny elements positioned in a precise location called swathe. It is 1x1 or 2x2 meters from the farm. Detectors are in ranch, and transmission among devices uses IEEE 802.15.4 procedure. The knowledge from machines is shipped to the hub using the isolated component. Learning communication employs radio permit methodologies. The gateway has a warehouse. It has a cloud stockpile. The technology is available to many customers. The software regains information from the detector. The outlines of the atmosphere and demand prerequisites display in the panel. Clients locate data about the existing situation using the board. The server keeps the facts regionally and then transmits the knowledge into the cyber world. It has two subcomponents. Learning accumulator is liable for deciphering atmosphere prerequisites using digital computers. The gateway component detains details from the detector.

The airborne detector medium (Link, Senner, & Claupein, 2013) uses an isolated administration flying mechanism. The system is E-Trainer 182. It has a diminutive robot pilot MP2028g. It has a universal aligning Strategy. The pilot navigates employing three-axis accelerometers, stress altimeter, and tension airspeed detectors. It has a wingspread of 1.8 m, a distance of 1.4 m, and a 4.2 kg weight. It has a giant brushless motor Kora 25 delivers 1 kW and an airscrew 125 inches. Fiberglass modifies and reinforces the manufacturer's machine mounting and the fuselage. Electrical fuel reserve for the motor use 5S 3.3 Ah Lithium–Polymer reserve. It disconnects from the energy store of the automatic pilot, gateway, telemetry, and detector. The independent aviation regulator supplies the opportunity to track paths within the domain. The navigated flights aim to flee above the exploratory area. The monolithic miniature spectrometer MMS1 does grouping. The component has a spectrometer accommodation set with an aberration-corrected hollow scratching, a thread cross-section converter as visual intake, and a junction rectifier collection.

The IoT-enabled intelligent agriculture system (Mahbub, 2020) has five corresponding subcomponents. They include perceiving, information examination, details transmission, presentation, and implementation units. The sensing component connects with the facts research segment. The unprocessed information is processed and investigated using various procedures. The system implements them in the microprocessor unit. It has two distinct circuits. The number of unconnected detectors is dispersed in the location. The device is equipped with necessary detecting elements. They gather data from the atmosphere and process them. The DHT11 is the climate sensor utilized to measure the climate conditions. PIR sensing device notices the movement of entities emitting infrared beacons to control the encroaching of living in the environment. nRF24L01 component is a recipient. The foremost device will assemble knowledge. The benchmark is set in the microcontroller of the head. The device corresponds the information of the detector with the threshold. The message is communicated to the client (when the benchmark reaches beyond the set value). The customer can observe the sensor feed using the web pages. The prominent device has the ability to robotic assignments. It can also inform the client with pH detector information. The agri-copter sprinkles and pollinates insecticide. ESP8266 component can access the ESP segment using a mobile.

The suggestion (Romanov, Gribova, Galelyuka, & Voronenko, 2015) surrounds the extensive territory of the yield. The microelectronic detectors with radio tracks locate on the greeneries of florae. The sensing element has a transceiver. This unit aims to collaborate devices with others deployed in the system. The components of the device are the information possession component, details computing segment, transceiver, and store subcomponent. GPS division, fuel producer, and mobilizer can also be part of the system. Knowledge management contains the evolution of detector medium, learning purchase, information condensation, facts accumulation, data computation, conclusion-making, and notice-making. Association of transmission among system devices has an adjustment of unconnected components, system management, preparation MAC-procedure, regional anatomy governance, the planning of grid connections, and institution of information distribution. The microcontroller has a 32-bit calibrator with 1–32 MHz clock pace, 2,4 GHz IEEE802.15.4 compliant transceiver, 4-input 10-bit ADC, and an inclusive combination of analog and digital attachments, etc. The JenNet-IP algorithm connects IEEE802.15.4-based unconnected web methodologies and the cyberspace procedure.

Figure 3.4 system implementation (Glória, Cardoso, & Sebastião, 2021).

The system (Glória, Cardoso, & Sebastião, 2021) has many detectors that gathers information, an actuator that controls the system, and an accumulator device that manages the incoming and outgoing data. LoRa radio component transmits the amassed information to the grouping device, a command panel, and a fuel reserve. The controller is accountable for appropriate actions. This device has a microcontroller, ESP32, a LoRa radio component, RFM95W, the climate port, and the agriculture drives. The microcontroller obtains the revised notification for farming and computes them to maintain the siphons. It attaches to a climate station, SEN 0186. It is qualified for organizing metric wind rate, breeze orientation, and drizzle. Machines obtain the information in real-time from the procedure. It pre-computes them and assigns them to the algorithm to comprehend. The methodology reeducates employing a new database. Figure 3.4 represents the same.

The information-gathering component (Kyaw & Ng, 2017) gathers facts employing five detectors. Water climate device collects water warmth of the fish reservoir. The water discharge pace instrument calculates water outpour pace from the fish chamber to yield setup. A computerized light detector enumerates the power of the surroundings. pH grade sensing device notices water pH status in the fish chamber. Ultrasonic ranger estimates the growth size. The warning division has a (green) LED glare, a red LED glow, and a beeper. This component shows a green glow when the approach is nutritious and transmits a red ray otherwise. The design adaptation division mechanically interferes and fixes the procedure anomaly by triggering the corresponding controller. The CPU defines the trigger or halt of the regulator. It decides w.r.t accumulated information and client predetermined rate. This component contains four regulators. The water boiler delivers an extra warmth origin when the water warmth drops underneath a beneficial degree. Subordinate water propels guarantees water stream from the fish chamber to yield in pump disaster. LED glare gives regular blue and red glow to promote yield increase when the ambient ray power penetrates a suffering condition. Fish tributary distributes fish food at the client's prescheduled period. The CPU has two branches. The first unit has an Arduino Mega, a Grove-Mega protection, and a relay panel. Arduino Mega has 54 input/output pins, and it corresponds with the detectors and controllers from the information gathering component and refinement segment. Grove Mega protected is on Arduino Mega to decrease the number of associations on the panel. The relay panel allows Arduino to handle the regulator by swapping the circuits on and off. The second unit has a Raspberry Pi 3 and a camera unit. Raspberry Pi is the management division. Camera component v2 allows a live sequence characteristic for Raspberry Pi. This camera segment can register high-definition tape after combining with Raspberry Pi. The online technology has Raspberry Pi to deliver GUI. The GUI depicts and approximates the live and documented detector rate. It permits the customer to watch the aquaponics installations and command the controllers. The portable technology uses an Android podium. It shows live detector readings and facilitates consumers to regulate the controller by employing assistance from the stockpile. The warehouse inducts transmission between the CPU and portable devices.

The automated estimation framework (Tripicchio, Satler, Dabisias, Ruffaldi, & Avizzano, 2015) has an airborne automobile, an RGB-D detector, and a programming medium. It is answerable for the steering stage and the post-calculation of the received information. It operates a retail low-cost RGB-D camera and an Asus Xtion Pro. It uses the component for the graphical investigation of the earth and employs a fixed-wing drone. The warehouse representative of the three-dimensional design of the domain accumulates geographical information for the texture examination of the various specializations. PID regulator uses perceiving methods, and a camera. The camera was not perpendicularly facing the terrain. The system does not notify height variation. It is required to spin the stockpile consequently to obtain a measurement. This activity is taken care in the acquisition phase. The principal segment Examination is performed in the investigation stage. The 3D corresponds to the PCA to discover the trajectory in the principal axis. The parameter will have maximum variance. The PCA swirl converts into a tested measurement domain utilizing Delaunay triangulation.

The work (Zecha, Link, & Claupein, 2013) specifies whether the method must be the virtual or offline approach. An offline process gathers information. It has different stages. They measure data followed by discovery, computation, and software. It supplies the opportunity to merge various origins of knowledge. A simulated strategy considers the metric detail in real-time for the conclusion accounting. It accomplishes by an assignment regulator, and a laptop. In mixture with Differential GPS, the information of the programs associates the information investigation and traceability.

The work (M Othman, Ishwarya, & Ganesan, 2021)uses three use cases. The first case grows lime and herbal growth. The second case has a farmstead farming herbal plant. The third case grew herb yield, hen, and mushroom. The work used portable devices, online applications, and necessary hardware components. The first feature forms and measures in the state of the authority container. This package operates the IoT and knowledge from the vegetables. The second segment investigates online technology. This component handles the authentic information from the widgets. This part enables an executive to use the methods for water needed for yield. The personnel view the notification. This knowledge undergoes examination to understand the crop requirements. The work also uses portable devices. The instrument supervises the working of the system.

There are three assignments to be completed in this approach (Bhanu, Rao, Ramesh, & Hussain, 2014). The knowledge assembles from the detector and the leader of the conservatory reports with the details at the hub using a user-friendly website and obtained messages. The devices calculate the climate and transmit these measures to the server. The server performs three functions. It corrects the mistakes in the intake information. The architecture consolidates the knowledge. The system saves the details. A user-friendly podium for the conservatory leader has technology.

The suggestion (Mekonnen, Namuduri, Burton, Sarwat, & Bhansali, 2019) is a dispersed sensor environment. It uses an open-source hardware podium, Arduino-based microcontroller, and ZigBee55 component to observe and manage requirements necessary to yield development as ground states, atmosphere, and climate necessities. The experiment is an off-grid photo-voltaic. It aggregates fuel and water information. The ranch acts on the dispersed unconnected detector. It observes and estimates different atmosphere restrictions. The architecture accumulates real-time climate knowledge to underrate surrounding influence and take more reasonable judgments to control utilities. The data assembled is unrestricted in the provincial and exterior datasets, and the clients can recover the details employing an involuntary portable technology. The mobile app permit customers to observe and communicate with the ranch framework. The system is fuel-efficient, price-effective, and permits the agriculturalists to supervise their farmstead. This system has a hub, six sensor networks, and a temperature console.

3.3.3 Environment monitoring

Atmospheric surveillance (Mukhopadhyay, 2012) (Acevedo, 2018) (Artiola, Brusseau, & Pepper, 2004) needs protection from harmful adulterants and pathogens released from various sources. Insecticides and weedkillers keep the soil and plants safe. The accidental discharges of other pollutants can derive from dripping tubes, subsurface repository tanks, garbage trash yards, and debris storage. The impurities can stay for many years and emigrate through large areas of the ground until they contact water sources. These prove to be a threat to living life. Figure 3.5 depicts EPA national primary drinking water standards for organic chemicals (Ho, Robinson, Miller, & Davis, 2005). Figure 3.6 represents EPA national primary drinking water standards for inorganic chemicals. (Ho, Robinson, Miller, & Davis, 2005).

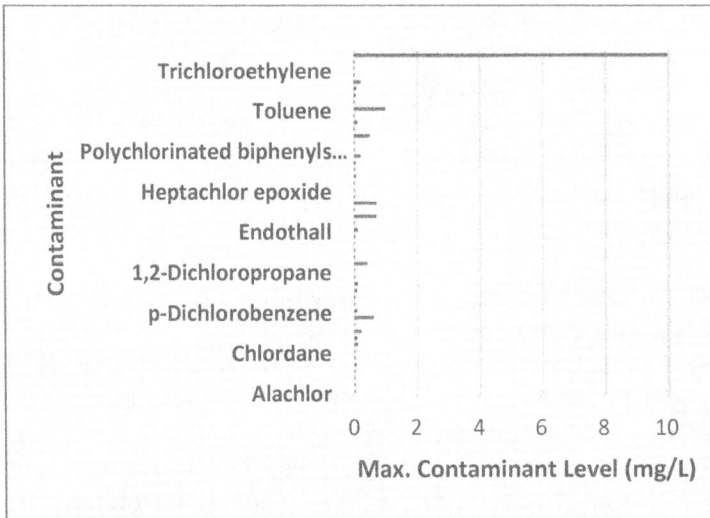

Figure 3.5 EPA national primary drinking water standards for organic chemicals (Ho, Robinson, Miller, & Davis, 2005).

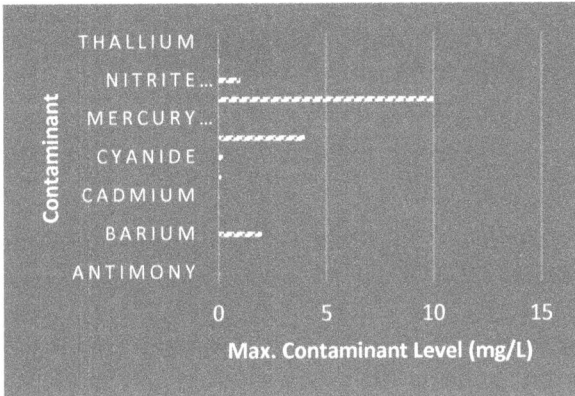

Figure 3.6 EPA national primary drinking water standards for inorganic chemicals. (Ho, Robinson, Miller, & Davis, 2005).

The Application tier (Barrenetxea, Ingelrest, Schaefer, & Vetterli, 2008) is accountable for accumulating information and later transmitting it to the server. Transport slab develops and acquires knowledge depending on the type of information parcels. The controlling information transmits to the next-door node. Two divisions contain the number of jumps required and another series number. The network tier ships packages to the MAC layer. The layer inserts the dispatcher identification and price of the path taken to the server. The steering conclusions constructs in this layer. The information admits by shipping a small container to the last stop. A simple back-off approach is employed to lessen crashes. SensorScope positions form using an aluminum structure fitted with a sun-oriented board, seven exterior detectors, (a hermetic) container, and surrounding electrical components. The system has Shockfish TinyNode2 having Texas Instruments MSP430 16-bit microcontroller, running at 8 MHz, and a Semtech XE1205 radio transceiver, operating in the 868 MHz band with a communication pace of 76 Kbps. It has 48 KB ROM and 10 KB RAM.

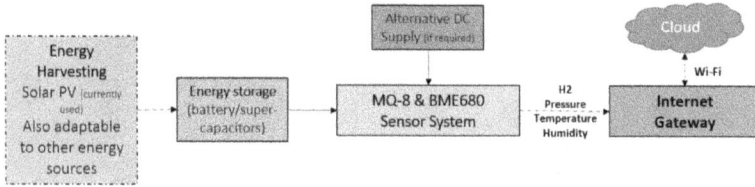

Figure 3.7 System block diagram for the sensing setup (Sunny, Zhao, Li, & Kanteh Sakiliba, 2021).

The detector method (Sunny, Zhao, Li, & Kanteh Sakiliba, 2021) combines numerous sensing devices with auxiliary models. It accomplishes real-time surveillance of the surrounding constraints. It furnishes precise dimensions for necessary limitations. The fuel garnering frame provides the design with a critical capability. It prolongs the lifespan of the procedure and decreases the requirement for keeping in long-term function. The unconnected transmission component delivers a link to the stockpile. It allows real-time observation and judgment-making with the help of synthetic intellect. The architecture employs two solar boards for fuel reaping with a declarable battery and an energy administration wiring, a hydrogen detector, a surroundings device BME680, a Wi-Fi Module ESP8266EX, and a Wemos organic light-emitting diode. The atmosphere information is in two formats. I2C buses transmit computerized knowledge. ESP8266E provides a clock beacon. The sequential learning Line is a bidirectional information sign. The analog facts utilize the ADC port of the microcontroller. ESP8266EX model transmits the information to the server. Figure 3.7 represents the same.

SnowFort (Liao, et al., 2014) is a framework to handle the need for a stretchable sensor podium. Execution necessities and standard technology limitations in structure and atmosphere surveillance storage are the configuration options for the procedure. The sturdiness and trustworthiness of the strategy are accomplished by the plan of the unconnected transmission procedure. Plainness and flexibility guarantee a single-hop system architecture. It is extendable by leaning on numerous servers. Intellect is delivered depending on high-performance computation and web systems to support visualization and data computation in real-time. The scheme supplies a high-level process to the system. APIs facilitate business growth. SnowFort seeks to be translucent. It disseminates as an open-source task. It can allow society the consequence of a sturdy, stable, dependable, adjustable, and uncomplicated answer for framework and atmosphere surveillance. The procedure framework has four components - an unconnected detector device, server, warehouse, and online medium. The element is the sensing division. The gateway intercedes a bunch of devices to construct a grid. The warehouse and the online podium form the conclusion aid procedure.

The architecture (Ullo, et al., 2018) has a construction of real-time benefits using a transportable gadget. The client can reserve dynamic help. The strategy will recognize journey tours considering the bottleneck. It will display different potential journey options with various prices and periods, incorporating other assistance in the region. The municipality has about 60,000 populace and average transportation. There are two diverse non-intrusive detectors embedded in the metropolitan region of Avellino. It collects atmospheric information and appropriate intoxication materials. The areas characterized by the city demonstrate pre-existing communication grids.

The architecture (Sun, Zhong, Jeong, & Yang, 2019) has three layers. The intake details kinds have time-series measures, vectors, and outcomes. No-SQL datasets InfluxDB stocks the supervising fact. The computing tier introduces the complex event processing motor. The system uses a Kafka environment. The finding stage is an online corroborating graphical investigation. Kafka has four stakeholders. The subject delivers a method of classifying transmissions. It functions as a medium of information storage for documents to be shared among entities. The issue information plan specifies by the customer for various detector classes. The case manages in several divisions for quicker knowledge recovery and information monotony. A Kafka creator pens a subject, and the buyer recites from a section. A Kafka agent is a hardware device in a dispersed approach that addresses the authentic reading, penmanship, and burden harmonizing. The customer is accountable for specifying the creator and client. Kafka delivers a technology medium for creators to make tailored designers and buyers. The Kafka associators authorize the composition of repositories that attach cases to known systems using traditional mediums. Business associators connect the framework to the information tier and understanding finding stage.

3.3.4 Smart homes

The three residences (Sprint, Cook, Fritz, & Schmitter-Edgecombe, 2016) are single-occupant flats. These houses are furnished with hybrid light with motion detectors on the roofs and door-climate sensing elements on cupboards and entrances. The devices uninterruptedly and modestly observe the day-to-day movements of the inhabitants by transmitting messages revises or instrument occurrences. CASAS-AR tags detector occurrences with workout titles in real-time as the happenings arise. The calculated occurrence distinguishes characteristics containing the sensing occurrence period, the dimensions of the descending activity, the affair computation for each detector within the duration, the span passed. Exercising information provides an exterior explanation. This description aids in generating the workout tags.

The suggestion (Dawadi, Cook, & Schmitter-Edgecombe, 2015) is intelligent residence-based conduct information analysis. It forecasts a person's routine fitness. The work assumes there is no association between an individual's fitness and everyday manners. The method observes the demeanor of a tenant employing clever dwelling devices. It measures their fitness utilizing traditional medical reviews. The work creates a procedure to foretell the mind and wellness inspection by creating the usefulness of real-world shrewd residence detector knowledge. The proposal suggests an experimental Inspection employing the Movement conduct practice to forecast the mental and dynamic scores. Clinical Assessment using Activity Behavior computes the workout-labeled detector knowledge to pull out exercise execution characteristics. CAAB removes numerical movement characteristics from the workout execution attributes to teach apparatus knowledge process that forecasts the mental and active scores. The detector learning gathers data from 18 real-world intelligent residences with (elder) inhabitants. An exercise identification procedure titles accumulated unprocessed information with the related exercises. CAAB employs facts gathered from the source without changing the tenant's behavior and surroundings. The strategy proposes a biologically reasonable approach to distinguish the ADL limitation and evaluate the mind and wellness of inhabitants.

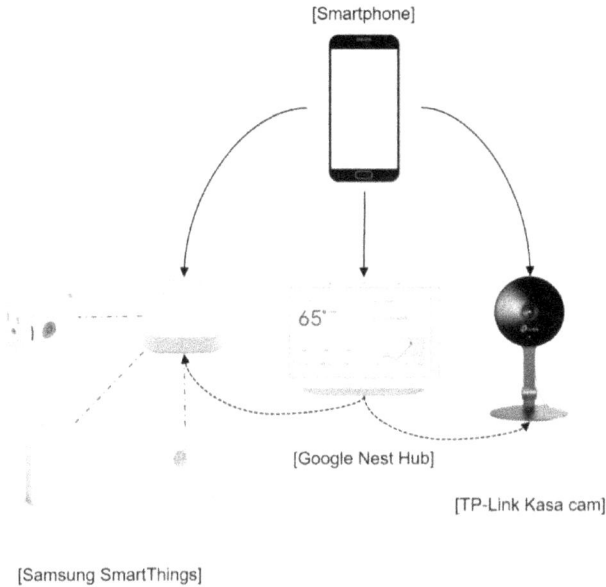

Figure 3.8 Smart home setup (Kim, Park, Lee, & Kim, 2020).

The work (Kim, Park, Lee, & Kim, 2020) begins with recognizing the precise procedures of an intelligent residence machine and executing investigations of stockpile. The operational sketches experimentations and develops research relics. The suggestion accumulates facts. The recommendation considers detailed mining approaches. The system later investigates the gathered information. The correlational research recognizes the associations between obtained knowledge and client conduct. The work operates examined knowledge in illegal examinations. The system includes a Google Nest gateway, a Kasa Camera, Detector, and a SmartThings Outlet. The server attaches to the low-end instruments. It functions as a hub for the information interaction between the residence appliances. The house baggage has an orator. The speaker regulates the system utilizing voice authorities. The outcome redirects to the camera. The gateway has a presentation unit. The component displays video or photos from the camera. The intake records the movements in the app on the mobile device. The portable device receives a notification on the camera sensing any motion. The Platform can manage the power. The Detector can sense the situation and oscillation in the system. Figure 3.8 represents the same.

Centre for Advanced Studies in Adaptive cluster (Gupta, McClatchey, & Caleb-Solly, 2020) contains a scope of tagged, partially marked, or unmarked exercise information, accumulated during a different time. The actions in this database are either written or unwritten. The suggestion concentrates on (routine) knowledge. Five shared learning satisfied these standards. The Aruba details have information of thirty-nine detectors. 34 are PIR detectors, and five are climate devices. The PIR instrument facts convey the tenant's biological activity in the surroundings. The work used 1,602,980 sensing occurrences out of 849,579. These are untagged data. The database had 11 activities. The work labeled these workouts.

The framework (Skubic, Alexander, Popescu, Rantz, & Keller, 2009) has six associates. The inactive physiological detector has an information supervisor and movement indicator, stove gauge, and bed finder. The occurrence-driven videotape device conceals recognizing attributes of the inhabitants. It has a logic motor that connects detector and tape knowledge. It investigates designs of conduct workouts. It has an adjustable signal administrator. It has a segment for delivering customization of device composition, beacon description, and details permit for the tenant. It has a knowledge host and online medium that supplies collaborating recovery and presentation. The system installs the setup in 17 flats. Accumulated facts are shipped to the datasets and retrieved using the host.

The wise residence surveillance (Suryadevara, Mukhopadhyay, Wang, & Rayudu, 2013) accumulates detector information and executes detailed examination. It notices conduct modifications of an aged. The home entities connect to simulated sensing components. It has various kinds of devices having electric and non-electric operations. They recognize the working of electrical appliances. They work on the discovery of existing associated entities to home-related things. The electric entities are microwave, water heater, roaster, chamber stove, and TV. The non-electric category objects are mattress, stool, lavatory, and couch. The system observes utilizing a Flexi Force sensor. The intake signs from the perception segments undergo combination and associates using radio message to the XBee component. The detector details of the entities send using XBee. The sensor network follows the IEEE standard 802.15.4 of ZigBee. The transmission administers Modem. The fuel store for the simulated component happens from an electric energy medium. The artificial segments are composed using mesh organization.

3.3.5 Smart cars

The new era is making intelligent systems a part of the environment. Intelligent vehicles (Vlacic, Parent, & Harashima, 2001) are one such trend that provides a wise solution to traffic congestion and surrounding conditions. The ability to witness automobiles is the feature of clever carrier administration. It authorizes us to collect details about the traffic situation and motorcar pace. The paragraph details the contributions made by various authors.

An investigation (Markevicius, et al., 2016) uses 24 different configurations of motors designed by various producers. The system examines the possibilities of using AMR sensing elements in transportation administration. The automobile visualizing architecture permits the measurement of the misshaping of the Earth's magnetic area generated by motorcars. The system has two detectors discovered at a length of 30 cm. The system measured every 1 cm along the X-axis path, apprehending the automobile location and the magnetic domain constraints. Examination approach of the X-axis was recited every 20 cm along Y-axis with the shifting of both AMR detectors. The information from the sensing element uses the RS 485 interface. The procedure of communicating knowledge and interpretation happens using the AM. Figure 3.9 represents the same.

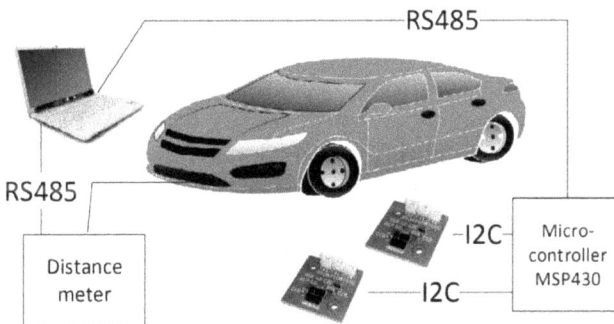

Figure 3.9 Experiment structure (Markevicius, et al., 2016).

3.3.6 Air pollution monitoring

Air pollution (Yi, et al., 2015) leads to diverse diseases like Heart disorder, Chronic Obstructive Pulmonary illness, stroke, and lung tumor. Populace breathing polluted air suffers from respiring, coughing, wheezing, and asthma. Hence the environment has to be monitored followed by adopting appropriate measures. To lessen the consequences of air contamination on mortal fitness, international atmosphere, and worldwide budget, administrations arrange extraordinary struggles on air contamination surveillance.

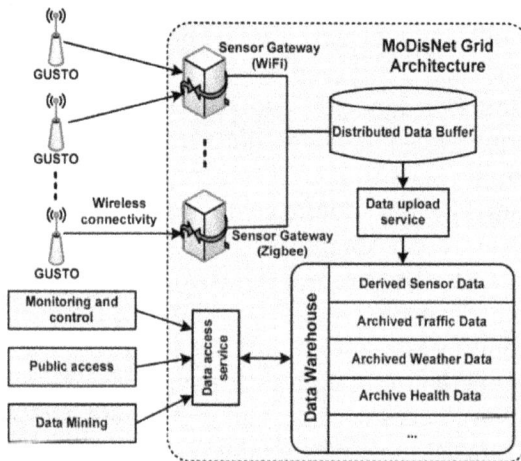

Figure 3.10 MoDisNet sensor grid architecture (Ma, Richards, Ghanem, Guo, & Hassard, 2008).

GUSTO detector (Ma, Richards, Ghanem, Guo, & Hassard, 2008) estimates contaminants with good precision and output at short intermissions. The system strains the knowledge and tallies data. The collection method uses technical procedures. MoDisNet Grid has adequate computational abilities to partake in the domain. It computes systematic information and transmits the same to its neighbors. The GUSTO elements link to the MoDisNet using hubs based on unconnected admission procedures. The tiny devices are competent in gathering the air contamination knowledge up to 1Hz frequency. It transmits the facts to the server using a multi-hop methodology. The hub associates the unconnected system with the IP spine. It watches the magnitudes of the information using the machines. It conducts the backpack optimization to detour handover crashes. The system allows repository admission using SQL databank. The catalogs stock and supports all the archived information. Figure 3.10 portrays the same.

The suggestion (Khedo, Perseedoss, & Mungur, 2010) is an unconnected detector grid. It is an air pollution surveillance system implemented in Mauritius. The component has two segments. The system recognizes information accumulated by the repository device using its credentials. The group leader gathers readings from machines and stocks them in an inventory. It investigates the items of the catalog using its identification. It eliminates the duplicates by using this methodology.

The recommendation (Tsujita, Yoshino, Ishida, & Moriizumi, 2005) uses a metal oxide gas detector to estimate nitrogen dioxide. The air is accepted using a membrane filter. The system uses SSWP01300 Millipore with a distance of 13 mm and aperture dimensions of 3m. The gush pace maintains 500 ml/min utilizing a needle vale. It watches them using an automatic discharge meter. The system has vapor detection in the second enclosure. It has a climate detector and a wetness machine in the first compartment. The two sections correlate with a sequence with all the sensing machines. The experiment considers four administrations surrounding surveillance positions close to the academy. The elements like sulfur dioxide, nitrogen dioxide, photochemical oxidants, and carbon monoxide are part of the measurement. The architecture also calculates meteorological parameters. The framework updates the mean of headings on an hourly basis over Cyberspace.

The architecture (Kadri, Yaacoub, Mushtaha, & Abu-Dayya, 2013)coordinates with four surveillance positions. The design has two subcomponents. A multi-gas observation system has many vaporous and climatological perception components. The information categorizes the unconnected transmission panel and the energy reserve method. The first two places have O3, NO2, and CO sensors, and the last two positions have H2S detectors. The smoke-identifying devices use nanotechnology semiconductors. It assesses the input by calculating the electrical conductivity of a delicate metal-oxide tier. The components associate analog intakes with the transmission panel. It uses Atmega 2560 microprocessor and accommodates an external MicroSD memory with 2 GB capacity. It accumulates details, calibrates them, and communicates. The panel has a GPRS dial-up connection

for unconnected links. The framework accepts a sampling of variables every 1 min. It computes and keeps the mean of five intakes. The
position connects using TCP/IP Internet link using GPRS dial-up link
with the M2M platform discovered at the backend gateway. Solar energy powers the system. The M2M medium is working on a backend
host. It encompasses the M2M transmission component, details vir-
tue element, a fact computing division, knowledge repository, and
forecast segment. The M2M transmission procedure serves over
GPRS or 3G grid and is accountable for joining all positions to share
knowledge.

3.3.7 Logistics

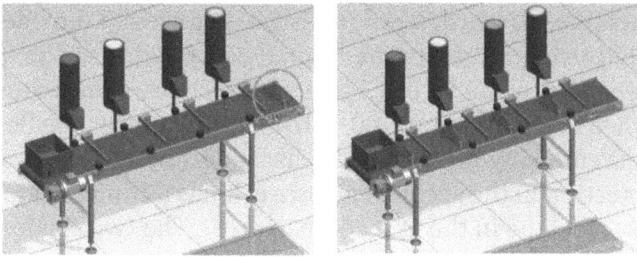

*Figure 3.11 (a) sensor position at the output of the line (b) Sensor posi-
tion at the workplace* (Vachálek, et al., 2021).

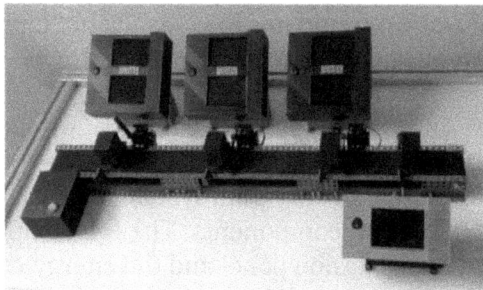

Figure 3.12 Physical model (Vachálek, et al., 2021).

The work (Vachálek, et al., 2021) is a robotic production system. It
permits the user to balance current or scheduled manufacturing connections utilizing a digital twin. The digital twin can explain the effect
of changes of different procedures on digital representatives. It can
forecast its consequence without the actual execution of designed

modifications. The association of real-time and digital setup uses the color of the detector to identify them. The framework mechanically creates the digital twin without exterior interference. It uses the intake of the stored datasets. The wired cyber-physical system substitutes the authentication algorithm. It employs four color-coded factory setups. Figure 3.11 represents the same. The detector recognizes the outcome using the technology. The workshop copies the invoice of the product considering its identifier. The framework describes material intake. It also intakes the present quantity of the yield. The photogate reviews the input. It exchanges a timer, which, based on the belt speed, initiates the gate and thus stops the component. The color detector senses the stop and start of the procedure. The example examines the relationship of the developed logistics approach to hardware and accumulates information for various operational trials. Figure 3.12 is the representation of the same.

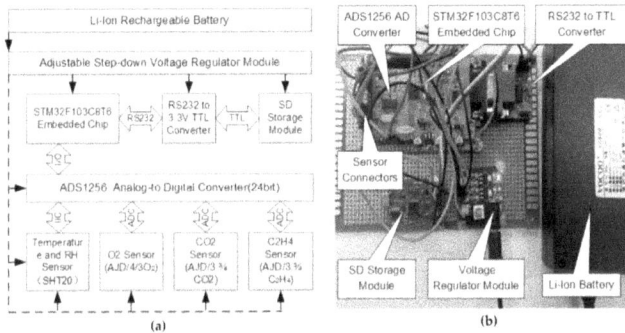

Figure 3.13 The hardware design of the multi-sensor monitoring detector. (Liu, Zhang, Li, Jemric, & Wang, 2019).

The Multi-sensor surveillance system (Liu, Zhang, Li, Jemric, & Wang, 2019) is a sequence Logistics information system. It is accountable for sleuthing, perceiving, and computation. The detectors compute warmth, wetness, the intensities of C2H4, CO2, and O2 steam. The resultant is a robotic indication. The outcome of vapor leads to an analog voltage sign. The beacons of smoke indicators are obtained by the analog portal of ADC. The climate detection machine attaches to STM32 by I2C interface circuit. ADC model transforms analog beacon into a robotic indication. It communicates using STM32. ADC transformation component has eight intake media and 24-bit exactness. The single-chip STM32F103C8T6 acquires, calibrates, and ships alerts at intermissions of 2 min. The authority chip in the switch representative utilizes the output. Clock chips regulate the duration of knowledge accumulation and make timing vibrations to trigger the reign chip. The details are transmitted to the warehouse using the serial pin. The RTC chip PCF8563 stores the data transmission details. The repository chip CH378Q saves the knowledge of sensors. Figure 3.13 is the representation of the same.

3.4 Use cases

Input	Prepossessing	Feature extraction	Supervised machine learning algorithm	Decision-level Fusion	Output
Finger Temperatures	prepossessing for FT	Individual features are extracted for each parameter using both time and frequency domains	One solution is received for each parameter using Case-based Reasoning (CBR)	A weighted similarity approach is applied in order to combine solutions from CBR cycle for each parameter	Final classification of the mental state in terms of stress or relax using Decision-level fusion
Oxygen saturation (SpO₂)	prepossessing for SpO2				
Respiration Rate (RR)	prepossessing for RR				
Carbon dioxide (CO₂)	prepossessing for CO₂				
Heat Rate (HR) from (ECG)	prepossessing for HR				

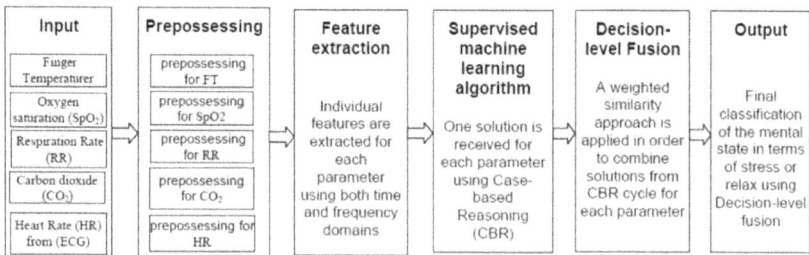

Figure 3.14 An overview of the classification scheme to identify mental state (Begum, Barua, & Ahmed, 2014).

The work (Begum, Barua, & Ahmed, 2014) accumulated 16 measures from eight personalities. Their epoch ranges from 26 to 50. The system notified players of the testing structure before the information assemblage. The information collected included age, hours of the resting night, and medicine consumption. The analysis happened in two phases. The psychophysiological tension of an individual is a six-division procedure. The experiment conducted route riding. The work instructed the populace to take a route and return to the destination. The components are removed from the detector's data using conventional characteristic mining methods. The five issues have five physiological signs. It recovers the identical issue from storage by corresponding to the signal. The procedure recovers five categories for five indications. The weighted likeness procedure delivers the final category. The gestures are fused employing a Multivariate Multiscale Entropy procedure. The recommendation removes components to create a new issue. The suggestion uses outcome in the CBR. Figure 2.8 portrays the same.

3.5 Challenges

3.5.1 Energy efficient clustering

The detector system (Yick, Mukherjee, & Ghosal, 2008) (Ambika. N, 2021) accumulates information and communicates the information to other nodes using a multi-hop methodology. The devices are small and contain small amounts of fuel. Hence the system has to consider some prevention measures to conserve energy and extend the lifespan of the nodes. The machine's design has to consider work diversity. The client must specifically describe the system implementation needs employing measures varying between dormancy, exactness, and dependability. The web must regard itself as a single commodity. It has to cooperate transmission procedures to eliminate monotonies in calculation and transmission and preserve an even spatial allotment of power.

3.5.2   Data gathering

Information accumulating procedures compose the system. They aid in gathering knowledge in the targeted atmosphere. The devices utilize various learning accumulating procedures to create an energy-efficient network.   The instruments are unconnected and widely spread in the environment. New approaches have to cope with the opposes like system interventions, device failures, and resource limitations. The investigators make trustworthy, price-effective, and non-invasive networks to watch the system. The gadgets have an identification aiding them to recognize themselves. A grid procedure cooperates to exude the fact from the labels.

3.5.3 Data analysis

The investigation (Heeringa, West, & Berglund, 2017) (Chambers & Skinner, 2003) portrays the efficiency of the applications in creating the procedures. The system uses these algorithms to handle and process various sensor knowledge. The methodologies use different mechanisms to extract significant designs from the datasets. The methods need to consider some resource limitations of the sensors. The constraints include calibration duration, memory, bandwidth utilization, and fuel balancing. The essential research indicates that the standard feature of the practice is the incorporation of precise procedures to handle various datasets.

3.5.4 Data storage

Traditional techniques (Mekki, Derigent, Rondeau, & Thomas, 2019) used in sensor networks need information transmitted from detector devices to an integrated server. Strategies like accumulation and reduction in knowledge communication lessen the price. These methods are essential for real-time or event-based methodologies. Functions use the query-and-aggregate procedure. Augmenting detector warehouse becomes essential over the period. The repository area is restricted, and transmission is costly. The depository prototype is crucial to help storehouse regulations and questioning needs.

3.6 Future directions

3.6.1 Healthcare applications

The system compels client needs and powers by current advancements in ironware and computer programs. The first epoch shows the possibility of changing the procedure for treatment. The struggle between reliability and confidentiality and the capability to install large-scale plans has to meet appliance conditions. It has to decide the period of unattached detector grids incorporated in wellness exercise and investigation.

3.6.2 Smart homes

The intelligent residence methodology delivers the possibility of substantial progress in the grade of life and tier of independence given to the populace. The expected outcome provisions the dwelling surroundings with clever and independent strategy. The system has centrally controlled networked machines. It supplies a mechanism for autonomous but secure occupancy. It constructs identification and surveillance movements of the everyday tenancy. The system can use the recent technologies to make things better.

3.6.3 Smart farming

Smart agriculture is the execution of different methodologies. It uses machines communicating using cyberspace. It stores the collected information on the stockpile. The system has to be made easier and economical to distribute it widely. The price has to encompass the labor price, cost to produce the yield. The framework has to be designed to provide better profit and yield. It should also encompass the effective use of water and provide good surveillance. The new technologies have to aid in bringing the changes towards betterment.

References

Acevedo, M. F. (2018). *Real-Time Environmental Monitoring: Sensors and Systems.* Boca Raton, Florida: CRC Press.

Alsiddiky, A., Awwad, W., Fouad, H., Hassanein, A. S., & Soliman, A. M. (2020). Priority-based data transmission using selective decision modes in wearable sensor based healthcare applications. *Computer Communications, 160*, 43-51.

Ambika, N. (2020). Reinstate Authentication of Nodes in Sensor Network. In S. N. Applications, *Sensor Network Methodologies for Smart Applications* (pp. 130-147). US: IGI Global.

Ambika, N. (2020). SYSLOC: Hybrid Key Generation in Sensor Network. In S. P., B. B., P. M., K. N., & H. W. (eds), *Handbook of Wireless Sensor Networks: Issues and Challenges in Current Scenario's. Advances in Intelligent Systems and Computing* (Vol. 1132, pp. 325-347). Cham: Springer.

Ambika, N. (2021). Wearable sensors for smart societies: a survey. In C. C. (eds.), *Green Technological Innovation for Sustainable Smart Societies* (pp. 21-37). Cham: Springer.

Artiola, J. F., Brusseau, M. L., & Pepper, I. L. (2004). *Environmental monitoring and characterization. .* Cambridge, Massachusetts: Academic Press.

Aydin, G., Hallac, I. R., & Karakus, B. (2015). Architecture and implementation of a scalable sensor data storage and analysis system using cloud computing and big data technologies. *Journal of Sensors.*

Barrenetxea, G., Ingelrest, F., Schaefer, G., & Vetterli, M. (2008). Wireless sensor networks for environmental monitoring: The sensorscope experience. *IEEE International Zurich Seminar on Communications* (pp. 98-101). Zurich, Switzerland: IEEE.

Benbasat, A. Y., Morris, S. J., & Paradiso, J. A. (2003). A wireless modular sensor architecture and its application in on-shoe gait analysis. *SENSORS. 2*, pp. 1086-1091. Toronto, ON, Canada: IEEE.

Bhanu, B. B., Rao, K. R., Ramesh, J. V., & Hussain, M. A. (2014). Agriculture field monitoring and analysis using wireless sensor networks for improving crop production. *Eleventh international conference on wireless and optical communications networks (WOCN)* (pp. 1-7). Vijayawada, India: IEEE.

Borgman, C. L., Wallis, J. C., Mayernik, M. S., & Pepe, A. (2007). Drowning in data: digital library architecture to support scientific use of embedded sensor networks. *7th ACM/IEEE-CS joint conference on Digital libraries* (pp. 269-277). Vancouver BC, Canada: ACM.

Capella, J. V., Perles, A., Bonastre, A., & Serrano, J. J. (2011). Historical building monitoring using an energy-efficient scalable wireless sensor network architecture. *Sensors, 11*(11), 10074-10093.

Chambers, R. L., & Skinner, C. J. (2003). *Analysis of survey data.* Hoboken, New Jersey: John Wiley & Sons.

Chiang, C. Y., Chen, K. H., Liu, K. C., Hsu, S. J., & Chan, C. T. (2017). Data collection and analysis using wearable sensors for monitoring knee range of motion after total knee arthroplasty. *Sensors, 17*(2), 418.

Cortés, R., Bonnaire, X., Marin, O., & Sens, P. (2015). Stream processing of healthcare sensor data: studying user traces to identify challenges from a big data perspective. *The 6th International Conference on Ambient Systems, Networks and Technologies (ANT-2015); the 5th International Conference on Sustainable Energy Information Technology (SEIT-2015). 52*, pp. 1004-1009. London, UK: ELSEVIER.

Dawadi, P. N., Cook, D. J., & Schmitter-Edgecombe, M. (2015). Automated cognitive health assessment from smart home-based behavior data. *IEEE journal of biomedical and health informatics, 20*(4), 1188-1194.

Dey, N., Ashour, A. S., Shi, F., Fong, S. J., & Sherratt, R. S. (2017). Developing residential wireless sensor networks for ECG healthcare monitoring. *IEEE Transactions on Consumer Electronics, 63*(4), 442-449.

Dias, G. M., Adame, T., Bellalta, B., & Oechsner, S. (2016). A self-managed architecture for sensor networks based on real time data analysis. *Future Technologies Conference (FTC)* (pp. 1297-1299). San Francisco, CA, USA: IEEE.

Díaz, S., Stephenson, J. B., & Labrador, M. A. (2020). Use of wearable sensor technology in gait, balance, and range of motion analysis. *Applied Sciences, 10*(1), 234.

Fidaleo, D. A., Nguyen, H. A., & Trivedi, M. (2004). The networked sensor tapestry (NeST) a privacy enhanced software architecture for interactive analysis of data in video-sensor networks. *2nd international workshop on Video surveillance & sensor networks* (pp. 46-53). New York, NY,USA: ACM.

Gajbhiye, P., & Mahajan, A. (2008). A survey of architecture and node deployment in wireless sensor network. *First International Conference on the Applications of Digital Information and Web Technologies (ICADIWT)* (pp. 426-430). Ostrava, Czech Republic: IEEE.

Ge, C., Yin, C., Liu, Z., Fang, L., Zhu, J., & Ling, H. (2020). A privacy preserve big data analysis system for wearable wireless sensor network. *Computers & Security, 96*, 101887.

Glória, A., Cardoso, J., & Sebastião, P. (2021). Sustainable irrigation system for farming supported by machine learning and real-time sensor data. *Sensors, 21*(9), 3079.

Gupta, P., McClatchey, R., & Caleb-Solly, P. (2020). Tracking changes in user activity from unlabelled smart home sensor data using unsupervised learning methods. *Neural Computing and Applications, 32*(16), 12351-12362.

Hariri, R. H., Fredericks, E. M., & Bowers, K. M. (2019). Uncertainty in big data analytics: survey, opportunities, and challenges. *Journal of Big Data, 6*(1), 1-16.

Heeringa, S. G., West, B. T., & Berglund, P. A. (2017). *Applied survey data analysis.* Boca Raton, Florida: chapman and hall/CRC.

Ho, C. K., Robinson, A., Miller, D. R., & Davis, M. J. (2005). Overview of sensors and needs for environmental monitoring. *Sensors, 5*(1), 4-37.

Kadri, A., Yaacoub, E., Mushtaha, M., & Abu-Dayya, A. (2013). Wireless sensor network for real-time air pollution monitoring. *1st international conference on communications, signal processing, and their applications (ICCSPA)* (pp. 1-5). Sharjah, United Arab Emirates: IEEE.

Khan, M., Silva, B. N., & Han, K. (2017). A web of things-based emerging sensor network architecture for smart control systems. *Sensors, 17*(2), 332.

Khedo, K. K., Perseedoss, R., & Mungur, A. (2010). A wireless sensor network air pollution monitoring system. *International Journal of Wireless & Mobile Networks*, 31-45.

Kim, S., Park, M., Lee, S., & Kim, J. (2020). Smart Home Forensics—Data Analysis of IoT Devices. *Electronics, 9*(8), 1215.

Kulkarni, P., & Ozturk, Y. (2011). mPHASiS: Mobile patient healthcare and sensor information system. *Journal of Network and Computer Applications, 34*(1), 402-417.

Kyaw, T. Y., & Ng, A. K. (2017). Smart aquaponics system for urban farming. *Energy Procedia,, 143*, 342-347.

Lai, D. T., Palaniswami, M., & Begg, R. (. (2011). *Healthcare sensor networks: challenges toward practical implementation.* Boca Raton, Florida: CRC Press.

Lee, S., Walker, R. M., Kim, Y., & Lee, H. (2021). Measurement of human walking movements by using a mobile health app: Motion sensor data analysis. *JMIR mHealth and uHealth, 9*(3), e24194.

Liao, Y., Mollineaux, M., Hsu, R., Bartlett, R., Singla, A., Raja, A., . . . Rajagopal, R. (2014). Snowfort: An open source wireless sensor network for data analytics in infrastructure and environmental monitoring. *IEEE Sensors Journal, 14*(12), 4253-4263.

Link, J., Senner, D., & Claupein, W. (2013). Developing and evaluating an aerial sensor platform (ASP) to collect multispectral data for deriving management decisions in precision farming. *Computers and electronics in agriculture, 94*, 20-28.

Liu, J., Zhang, X., Li, Z. Z., Jemric, T., & Wang, X. (2019). Quality monitoring and analysis of Xinjiang 'Korla'fragrant pear in cold chain logistics and home storage with multi-sensor technology. *Applied Sciences, 9*(18), 3895.

Lo, B. P., Thiemjarus, S., King, R., & Yang, G. Z. (2005). Body sensor network–a wireless sensor platform for pervasive healthcare monitoring. *3rd Int. Conf. Pervasive Computing* (pp. 77-80). Munich, Germany: IEEE.

M Othman, M., Ishwarya, K. R., & Ganesan, M. (2021). A Study on Data Analysis and Electronic Application for the Growth of Smart Farming. *Alinteri Journal of Agriculture Sciences, 36*(1), 209-218.

Ma, Y., Richards, M., Ghanem, M., Guo, Y., & Hassard, J. (2008). Air pollution monitoring and mining based on sensor grid in London. *Sensors, 8*(6), 3601-3623.

Mahbub, M. (2020). A smart farming concept based on smart embedded electronics, internet of things and wireless sensor network. *Internet of Things, 9*, 100161.

Markevicius, V., Navikas, D., Zilys, M., Andriukaitis, D., Valinevicius, A., & Cepenas, M. (2016). Dynamic vehicle detection via the use of magnetic field sensors. *Sensors, 16*(1), 78.

Mekonnen, Y., Namuduri, S., Burton, L., Sarwat, A., & Bhansali, S. (2019). Machine learning techniques in wireless sensor network based precision agriculture. *Journal of the Electrochemical Society, 167*(3), 037522.

Mrabet, H., Belguith, S., Alhomoud, A., & Jemai, A. (2020). A survey of IoT security based on a layered architecture of sensing and data analysis. *Sensors, 20*(13), 3625.

Mukhopadhyay, S. C. (2012). *Smart sensing technology for agriculture and environmental monitoring.* Berlin Heidelberg.: Springer.

Nittel, S. (2009). A survey of geosensor networks: Advances in dynamic environmental monitoring. *Sensors, 9*(7), 5664-5678.

Raghavendra, S., C., Sivalingam, K. M., & Znati, T. (2006.). *Wireless sensor networks.* cham: Springer.

Romanov, V., Gribova, V., Galelyuka, I., & Voronenko. (2015). Multilevel sensor networks for precision farming and environmental monitoring. *International Journal Information Technologies & Knowledge, 9*(1), 3-10.

Shah, R. C., Roy, S., Jain, S., & Brunette, W. (2003). Data mules: Modeling and analysis of a three-tier architecture for sparse sensor networks. *Ad Hoc Networks, 1*(2-3), 215-233.

Skubic, M., Alexander, G., Popescu, M., Rantz, M., & Keller, J. (2009). A smart home application to eldercare: Current status and lessons learned. *Technology and Health Care, 17*(3), 183-201.

Sprint, G., Cook, D., Fritz, R., & Schmitter-Edgecombe, M. (2016). Detecting health and behavior change by analyzing smart home sensor data. *IEEE International Conference on Smart Computing (SMARTCOMP)* (pp. 1-3). St. Louis, MO, USA: IEEE.

Stripelis, D., Ambite, J. L., Chiang, Y. Y., Eckel, S. P., & Habre, R. (2017). A scalable data integration and analysis architecture for sensor data of pediatric asthma. *33rd International Conference on Data Engineering* (pp. 1407-1408). San Diego, CA, USA: IEEE.

Suakanto, S., Engel, V. J., Hutagalung, M., & Angela, D. (2016). Sensor networks data acquisition and task management for decision support of smart farming. . *International Conference on*

Information Technology Systems and Innovation (ICITSI) (pp. 1-5). Bandung, Indonesia: IEEE.

Sun, A. Y., Zhong, Z., Jeong, H., & Yang, Q. (2019). Building complex event processing capability for intelligent environmental monitoring. *Environmental modelling & software, 116*, 1-6.

Suryadevara, N. K., Mukhopadhyay, S. C., Wang, R., & Rayudu, R. K. (2013). Forecasting the behavior of an elderly using wireless sensors data in a smart home. *Engineering Applications of Artificial Intelligence, 26*(10), 2641-2652.

Tripicchio, P., Satler, M., Dabisias, G., Ruffaldi, E., & Avizzano, C. A. (2015). Towards smart farming and sustainable agriculture with drones. *International Conference on Intelligent Environments* (pp. 140-143). Prague, Czech Republic: IEEE.

Tsujita, W., Yoshino, A., Ishida, H., & Moriizumi, T. (2005). Gas sensor network for air-pollution monitoring. *Sensors and Actuators B: Chemical, 110*(2), 304-311.

Ullo, S., Gallo, M., Palmieri, G., Amenta, P., Russo, M., Romano, G., . . . De Angelis, M. (2018). Application of wireless sensor networks to environmental monitoring for sustainable mobility. *International Conference on Environmental Engineering* (pp. 1-7). Milan, Italy: IEEE.

Vachálek, J., Šišmišová, D., Vašek, P., Fiťka, I., Slovák, J., & Šimovec, M. (2021). Design and implementation of universal cyber-physical model for testing logistic control algorithms of production line's digital twin by using color sensor. *Sensors, 21*(5), 1842.

Venkatramanan, V., Shah, S., & Prasad, R. (2020). *Global climate change: resilient and smart agriculture.* Cham: Springer.

Vlacic, L., Parent, M., & Harashima, F. (2001). *Intelligent vehicle technologies.* Amsterdam, Netherlands: Elsevier.

Yan, H., Huo, H., Xu, Y., & Gidlund, M. (2010). Wireless sensor network based E-health system-implementation and experimental results. *IEEE Transactions on Consumer Electronics, 56*(4), 2288-2295.

Yang, G. (2006). *Body sensor networks* (Vol. 1). London: Springer.

Yi, W. Y., Lo, K. M., Mak, T., Leung, K. S., Leung, Y., & Meng, M. L. (2015). A survey of wireless sensor network based air pollution monitoring systems. *Sensors, 15*(12), 31392-31427.

Yick, J., Mukherjee, B., & Ghosal, D. (2008). Wireless sensor network survey. *Computer networks, 52*(12), 2292-2330.

Yick, J., Mukherjee, B., & Ghosal, D. (2008). Wireless sensor network survey. *Computer networks, 52*(12), 2292-2330.

Zecha, C. W., Link, J., & Claupein, W. (2013). Mobile sensor platforms: Categorisation and research applications in precision farming. *Journal of Sensors and Sensor Systems, 2*(1), 51-72.

4

RFID AND DATA ANALYTICS

Abstract

RFID patches are tiny, radioed mechanisms that serve to recognize things and characters. It is a method of explicitly naming gadgets to promote by measuring machines. The technology got its potential when retailing titan Wal-Mart announced for best 100 suppliers to provide RFID-enabled purchases by January 2005. The labels are testimony representation. The chapter details the working of Radiofrequency Identification technology. The methodology is used widely in many applications – healthcare, logistics, and supply chain management.

4.1 Introduction to RFID

Radiofrequency Identification (Want, 2006)(Ahuja & Potti, 2010) is abbreviated as RFID. It is a methodology that has moved from uncertainty into significant reinforcements. It helps expedite the administration of produced assets and supplies. RFID replaces barcode methodology. It is being autonomous of the route of vision intricacies and examining things from a horizon. It gives the decreased activity levels, increased distinctness, and enhanced record administration. It permits licenses from a range. The labels maintain a broader assortment of individual IDs. It can accommodate supplementary information such as entrepreneur, commodity class, and measured surrounding constituents such as warmth. The arrangements can recognize many various labels established in the corresponding common field without individual support. They have an unlimited shell life. It can get into a working adhering design. An inactive fragment has following components- an antenna, a semiconductor microchip connected to the antenna. The reader is responsible for powering and communicating with a tag. The label transmitter captures power and transfers the tag's ID. The encapsulation maintains the tag's integrity and protects the antenna and chip from environmental conditions or reagents. The encapsulation could be a small glass vial or a laminar plastic substrate with adhesive on one side to enable easy attachment to goods.

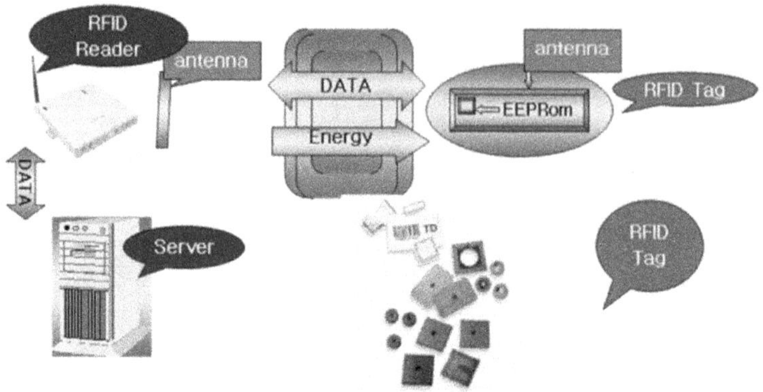

Figure 4.1 RFID System Structure (Park & Eom, 2011)

The first title (Park & Eom, 2011) attaches to the object. It traces the same. The reader has various commitments counting fueling the label, recognizing it, understanding information obtained from it, reporting to it, and transmitting with a knowledge assemblage. The cluster component gathers details from the reader, documents the learning into a warehouse, and delivers entry to the facts in many conditions. An RFID design conveys electromagnetic waves. Figure 4.1 portrays the same.

Two fundamentally different RFID design approaches exist for transferring power from the reader to the tag. The two categories are:

4.1.1 Active labels

They have their energy reservoir. It conveys a more powerful beacon, and users can reach them from a farther distance. The onboard strength origin is outsized and more valuable. Hence dynamic RFID operations operate best on extensive parts followed over great ranges. Low-power dynamic chips are usually imperceptibly more extensive than tarots of playing passes. They reside asleep till they appear in the span of a handset or can continually announce a beacon. They function at more important recurrences 455 MHz, 2.45 GHz, or 5.8 GHz. It depends on the application read area and retention obligations. Reading devices can interact with dynamic RFID patches from 20 to 100 measures. The system structure (Al-Ali, Aloul, Aji, Al-Zarouni, & Fakhro, 2008) has a connection operator that controls all interface operations performed at the mentor service. The mediator manages and interprets the knowledge. The beneficial GUI has a store of extracted data and user learning. The framework uses RF active reading components. It communicates with the primary posting using three various interfaces - serial wire, LAN line, and WLAN transmitter. The construction uses RF labels with a built-in reservoir. It can trail children in community parks, playgrounds, shopping centers, etc. It considers that the structure is to be used to follow kids' mobility in Dubai Global Village. It is a global presentation that magnetizes the populace from several nations. Relations with their ward's appointment the township at the level of 40000 to 50000 company per date. The town safety workplace gets hundreds of vanished teenager cases daily. The defense executives will have a rigid instance to situate the missing offspring.

Figure 4.2 is the picture of RFID.

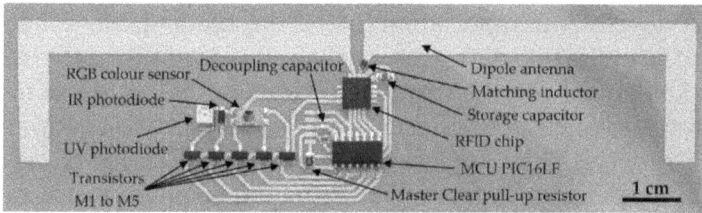

Figure 4.2 diagram of RFID tags (Escobedo, et al., 2016)

4.1.2 Passive labels

Inactive chips are very cheap. They cost 20 cents apiece. It contains about 2 Kbits of retention. The reading part can continually transmit its beacon or disseminate it on request. When a component is within the reader's reach, it sustains an electromagnetic beacon from the reader within the tag's antenna. The label stores the power from the sign in an onboard capacitor, a method called inductive coupling. When the capacitor has established up sufficient charge, it can power the RFID tag's systems. It sends a vibrated flag to the reading component. The acknowledgment beacon comprises the knowledge saved in the ticket. The connection between the translation segment and inactive piece uses a pair of methodologies to accentuate the ID flag. Low-frequency is smaller than 100 MHz. It passes learning by delivering strength from the capacitor to the chip loops in alternating intensities over the period. It changes the communication pulse released by the patch. The interpretation system recognizes these diverse fluctuations and can practice these modifications to demodulate the regulations. The higher frequency is more numerous than 100 MHz. It carries the beacon using backscatter, in which the tag's course becomes the endurance of the tag's antenna. This diversity in impedance causes the broadcast of RF fluctuations. The interpreter can choose and demodulate. It performs at regularities of 128 kHz, 13.6 MHz, 915 MHz, or 2.45 GHz.

4.2 RFID Architecture

The recommended monitoring Arrangement (Chen, Chen, Chen, & Chang, 2007) keeps preservation and supervision of the knowledge. The various commands for particular kind of traced things define the functions of rolling gadgets. The source links to the IP system. The hub advises the associations with the reading component. It watches the indication of triggered labels. It controls the activities of marker habits and registers suitable ticket learning to the confined listing core installed in their area. The reading part will continually produce reports. They endlessly observe the progress of the description. The created communication by the component assembles at the focal point. The center examines the post to analyze the performance of labels. It cleans additional messages and creates reports retaining necessary data. It controls the knowledge of gadgets in a particular area. The Global Registering center manages the elasticity of all local ones. The framework is a dual-tier structure with the global center doing the enrollment.

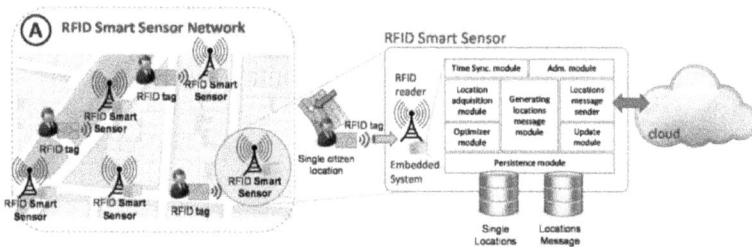

Figure 4.3 Architecture of RFID (Mora-Mora, Gilart-Iglesias, Gil, & Sirvent-Llamas, 2015).

The work *(Mora-Mora, Gilart-Iglesias, Gil, & Sirvent-Llamas, 2015)* is a processor designed that discovers the progress of characters in the town. It integrates various new methodologies and answers of IoT to create a complete practice to trace and follow residents in the area serving circumstances. The centralized segment obtains the knowledge from devices. They contribute resident tracing assistance to assist when producing judgments on modern arrangement to the users. The method is the procurement of places of civilians in complex and unconstrained conditions. It facilitates the clear alliance of residents into the practice. Rule administration is a procedure for structuring a complicated method into a series of jobs. It displays claims that convert information into some other producing components. The sub-process study is resident's area procurement. It focuses on learning technology foundations to get subject knowledge. The system of interaction and structuring involves securing the accurate response of the occupant position knowledge. The course focuses on providing official national movements to third individuals. The RFID sensing devices consist of reading devices with computation and transfer abilities. The centralized store-based practice focuses on the construction and providing the citizens with a translucent style. The method offers a recommendation design. It is the amalgamation of designs and software composition systems. The detector is bartered by the RFID Controls system. It is a configurable and adaptable device with pair of antennas. It has the highest reading pace of 400 tags. The best interpretation strength is 1 W. This tool provides a link to two antennas of two 50 Ohm MCX connectors with 6.5 dB gain. It provides orbicular polarization and a diffraction decoration with 60°/60° beam thickness. The suggested structure is the Enterprise Service Bus combination foundation. The software scheme designs defined in the structural design achieve a stand. The work implements two Mule ESB service flows to achieve the place communication attainment Service and the stream production Service. Figure 4.3 portrays the same.

4.3 Characteristics of RFID

RFID is a field of computerized association. It is a consideration to develop as pervasive computation processing. It has similar thought to barcoding. It improves information computation. It is equivalent to enduring methodologies. The characteristics are:

4.3.1 Operating frequency

It uses an extensive category of purposes differing from the structure admittance command (closeness tickets) to furnish connection tracking, duty gathering, transportation parking entree administration, local assets supervision, following books facility, fraud restriction, transportation immobilizer operations, and track circling commodity association and progress tracking. RFID labels comprise two chief components. The combined course consists of the microprocessor, memory, and an antenna. The purpose of the antenna is to determine the scope of the ticket. RFID badges classify into a couple of classes depending on their information warehouse ability. Most Read-Only labels do not know accommodation space. They only have an individual ID prewritten to them. It guides them to storage. It provides learning about the gadget. RFID tickets recognize dynamic and inactive status. Inactive checks depend on the electromagnetic domain produced by the RFID reading device to get initiated. Dynamic labels have built-in batterings. The tickets do not depend on the electromagnetic discipline of the reader to get stimulated. It expands the scope of the practice.

Various RFID practices perform at a mixture of communications wavelengths. Each spectrum of cycles allows its driving reach, strength demands, and production. The procedures limit themselves to complex commands or limitations that restrict w.r.t purposes. The working incidence manages which concrete elements generate RF beacons. Minerals and fluids impersonate the difficulty in tradition.

4.3.2 Method of coupling

RFID reading/writing component conveys to the tags or teams with RFID labels employing diverse methods. It includes backscatter, inductive and capacitive coupling.

4.3.3 Transmission range

The system is an assortment of services, varying from shared conveyance to animal tag to outcome search. RFID markers follow components reasonable for industrialization and logistics objectives.

4.3.4 Data storage capacity

The RFID transponder labels deliver a notable serial digit. It can keep up to 116 bytes of client information. It permits only approved admission. The NFC qualified appliance serves as a vigorous reader and a writer.

4.3.5 Power supply

RFID has AC to DC energy converter. It transforms the alternating current from the doorway to direct electrical energy.

4.3.6 Read only/read-write

RFID read/write heads are functional as near and far-field elements. The command components administer the activity. The read-only regulations and inconsistent information sites also aid in managing the operations. They can compose datasets.

4.4 Applications

4.4.1 Surveillance

The developed search strategies incorporate the present technologies of CCTV video administration and RFID search mechanisms resulting in an observable capacity to outline classified support.

The analysis situation (Golding & Tennant, 2008) investigates educational archives. It is a fully executed RFID operation and has approximately 120,000 volumes. The index browser and the self-check service are part of the investigation. The examination specimen consisted of 200 publications- paperback, hardcover, and element adhesive of differing dimensions. The reads were performed and registered on 40 occasions. An investigation is on the nature of titles that encounter reading problems. Based on each test, an assumption was considered that communicate to analysis puzzles noted beginning. The beacons broadcasted from the browser were significantly employed by wood-based on strength levels reported by the analyzer with a translation of -68.51 dbm. The alloy cupboards incorporated less at -56.42 dbm. The presentations register had the tagged book on ridges. The consequences for timber and element mantelpieces were -65.24 dbm and -61.98 dbm sequentially. The examinations record that the variance in the energy level between both shelving types with and without chapters was not meaningful. The spectrum analyzer manages a reiteration scan of the RFID browser to establish the frequency of service at 13.56 MHz. Ventures were reviewed and seen in simulcast groups (97 – 107MHz), 800MHz cellular combos, and 2.4GHz links.

4.4.2 Healthcare

An RFID strip wraps about the wrist of a sick. The infirmary holds a trail of suffering. The wristband maintains all applicable information such as name, therapeutic history, aversions, etc. RFID can enable the physician to know the sick's medical conditions.

The video sensing device system (Joshi, Acharya, Kim, Kim, & Kim, 2014)is for aging surveillance. The RFID device usage considers the communication and response coverage capability of the gadget. Chips are present on medications, couches, carpets, showers, and belongings of aged sufferers. The sick equips RFID labels using suitable RFID chips in combination with warmth and displacement sensing tools based on their physical and psychic well-being health, and guardians are with wearable portable RFID tools. The medications have inactive RFID badges. The knowledge is fixed or movable in devices reach. It confines to server or gateway through an electrified or radio system. The host contains an elderly management system. The restricted host relays communications to the telecast assembly to transfer signals, suggestions, or distress communications to experts, guardians to take quick response. The protection and organizational objectives, knowledge is collected in the central place host. It also demands the simulcast system to transmit communications to all destination beneficiaries. The victims having mobile RFID chips can interact in a multi-hop method when the RFID reading component is out of the response area. When an aged have to go outdoor, the RFID interpretation at the door scans patches connected to all things stocked by the sufferer. Figure 4.4 represents the same.

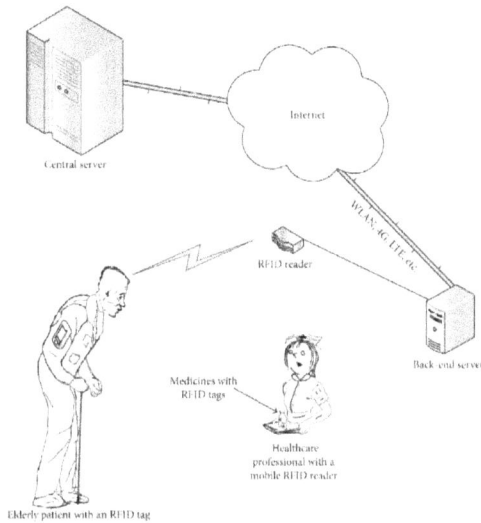

Figure 4.4 Components of RFID system (Joshi, Acharya, Kim, Kim, & Kim, 2014).

The work is an RFID-based ecosystem (Amendola, Lodato, Manzari, Occhiuzzi, & Marrocco, 2014) promoting innovative pervasive healthcare settings. It could be a Smart-House provided with a dispersed arrangement of collections, implementing a uniform and sturdy coverage in the most appropriate locations. It is a complex set of battery-less fragments with sensing ability. Ambient sensing devices can recognize environmental parameters of the atmosphere like warmth, moisture. The presence of dangerous agents will authorize to quantify the wellness of the situation. It analyzes its association with the person's well-being status. Wearable and implantable chips sanction to give secondary knowledge about the appearance of a physique inside a studio. The movement, the communication with characters, and futuristic body-centric utilization exhibit data about the fitness status of the prosthesis or unnatural glands. Data accumulated from the RFID ecosystem utilize data mining procedures. It examines the changes and trajectories of the client. It classifies its indications every day, and hibernation recognizes dangerous situations and emits warnings. The technologic foundation has to be submissive with the electromagnetic security model concerning the energy consumption in the individual anatomy and the transmitted course intensity in dwelling surroundings.

It is a ((Lee & Shim, 2007) model predicting the likelihood of adopting RFID within organizations in the healthcare industry. An organizational RFID adoption model is proposed and empirically tested by a survey using a sample of 126 senior executives in U.S. hospitals. The model posits that three categories of factors, technology push, pull, and presence of champions, determine the likelihood of adopting RFID within organizations. This study also found that the relationships between those three categories and the likelihood of adopting RFID are strengthened or weakened by organizational readiness. The present study includes one important dimension of the decision maker-presence of champions. A web-based survey was used to collect data for this study. There are well-documented practical problems with the paper-based form of data collection including poor response rates, slow response, and manual transcription of data from a hard copy questionnaire to an appropriate statistical analysis tool. The validity of the structural model of the research model was tested using LISREL 8.72. The fit statistics indicate that the research model provides a good fit to the data. The findings indicated that need-pull, technology-push, and the presence of champions strongly impact on the likelihood of adopting RFID in hospitals.

4.4.3 Optimizing the work flow

The gush of knowledge between store chain associates is a tactical action that improves implementation. The swapping and intercommunicating details boost the functioning of the system. The incorporation of facts in a shared channel implicates many workouts - the sharing of yield knowledge, merchandise status, distribution, cargo, accommodation, deals, and execution within businesses and among the system associates.

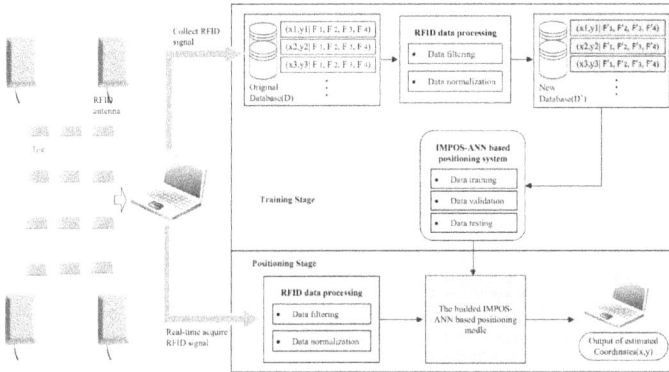

Figure 4.5 The framework of the PSO-ANN-based indoor positioning system (Wang, Shi, & Wu, 2017).

RFID location network (Wang, Shi, & Wu, 2017) utilizes an enhanced swarm optimization procedure to adjust the feed-forward neural network. The technique can understand the connection between the received signal strength indication and label places. The status of a new title can calculate using the designated location standard. It operates received signal strength indication values estimated by the RFID reader and the area of labels. The knowledge aspect of the asynchronous transition plan is created, which creates the education characteristic dynamically modified in the repetitive function for the program. The Gaussian sieve executes the received signal strength indication values. The simulation utilizes the Ackley function and the Rastrigin function. Figure 4.5 pictures the same. Table 4.1 summarizes the contribution towards optimizing workflow using RFID.

Table 4.1 contribution towards optimizing workflow using RFID

Contribution	Description	Experiment specification
(Faschinger, Savtry, Patel, & Tas, 2007)	• The objects are RFID-labelled and haphazardly positioned in racks that have RFID readers. • It reads the RFID markers of particular things and transmits them to an information warehouse. • Ant-colony-based augment procedure to compute the host path that a client ought to heed from his existing place to the different racks of needed objects.	• The system uses IEEE 802.15.4 radio transceiver.
(Poisai Arunachalam, et al., 2017)	• Mayo Clinic operates a customized real-time panel that provides real-time remarks of sick. • The position-time coordinating system considers two distinct longsuffering and their relations with a doctor and nursemaid in real-time.	• The setup has 750 RFID sensors. • It has 75 beds on 1.25 acres.
(Xie, Shi, & Issa, 2011)	• The RFID fragment has incorporated wiring for holding and computing knowledge, moderating and intensifying a radio-frequency (RF) sign. • The scanner reviews components of the steel skeleton with an RFID barcode. • It decodes into a regulation that defines the steel • The information transmits to the transmission approach and alters the shade of the established steel associate in the 4D VR prototype.	• The framework uses AutoCAD, SDS/2, and MS Project. • It combines Building Information Modelling (BIM), Radio Frequency Identification (RFID), and Virtual Reality (VR) Simulation.
(Peng, Liu, He, Yu, & Li, 2019)	• The sophisticated occurrence computation is RFID-enabled retail supply administration. • The port entrance RFID reader catches the occurrence and incorporates label (When yields deliver to the retail mart using the dock entry) • The RFID strategy sieves RFID readings and revises the supply details in the judgment method. • The various kinds of occurrence questionary define vendor stockpile workflow activity over the RFID information on numerous tagging classes.	• The system has 12 event types. • It has 100 – 500 kinds of products. • It is implemented with C# on a laptop with 4 GB memory and two 2.6 GHz processors.

4.4.4 Smart Environment

The technology is unconnected and aims to send information. It recognizes and follows labels connected to things. Knowledge kept on RFID titles can be modified, revised, and sealed. The identification has an antenna accountable for delivering transmission. The reader includes a trans-receiver that yields a vibration of electromagnetic waves. The transponder acquires the communication later corrected to obtain the dc energy store for the IC recollection.

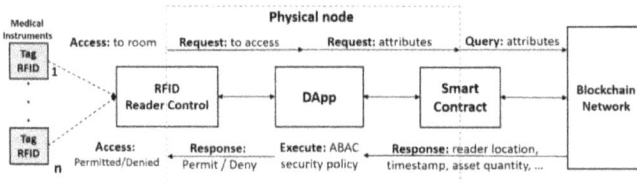

Figure 4.6 Decentralized system architecture based on Ethereum (ETH) blockchain (Figueroa, Añorga, & Arrizabalaga, 2019).

The physical device (Figueroa, Añorga, & Arrizabalaga, 2019) has RFID Reader administration, decentralized application, and intelligent agreement. The RFID-RC transmits a demand for admission to the application when a therapeutic tool tries to earn entry to a chamber. The application ships a question using an agreement to the blockchain system. It yields characteristics associated with the repository. It employs the features to manage the access control model based on the attribute's protection approach. It defines whether label entry is approved or rejected. The admission management reviews the topic's characteristics, access control guidelines, entity's qualities, and surrounding states. Figure 4.6 portrays the same. Table 4.2 summarizes the contribution towards RFID in Smart Environments.

Table 4.2 contribution towards RFID in smart environments

Contribution	Description of work	Experiment specification
(Jeykumar & Blessy, 2014)	• The framework utilizes clever bar codes. It communicates with the architecture to trace the produce stored in the cart. • The labels intercommunicate with the reader to recognize the object. • The component sends data to the dealer and factories. • The client's bank sends a notification on the amount deduction.	• MIT's Auto-ID Center is working on an Electronic Product Code (EPC) identifier • smart tag has 96 bits of knowledge (serial number, yield producer details) • Serial number is of 40-bit
(Bouchard, Fortin-Simard, Gaboury, Bouchard, & Bouzouane, 2014)	• The system positions the label in a particular area. It gives a standard to pinpoint another one. • Repetition is a fixed period intermission that achieves localization. • It decreases the portion of wavering by using a Gaussian mean to the RSSI obtained by the antennas.	• A-PATCH-0025 antennas is used. • Two counters of 170 × 60 cm and 129 × 60 cm are used. • Area is divided into 64 square zones of 30 cm per 30 cm
(Baeg, Park, Koh, Park, & Baeg, 2007)	• The system commences the cleaning after the owner leaves his residence. • It recognizes and brings a summoned entity. • The clever assistance intercommunicates with machines. • Unexpected events like detecting intruders are reported to the home server. The robot captures the picture and transmits it to the predefined location (On command from the server).	• ZigBee protocol and EPCglobal Gen2 is used • 32-bit ARM-based processor detects the status of smart items
(Park, Baeg, Koh, Park, & Baeg, 2007)	• The system mines graphical descriptors and stocks into the warehouse after enrolment. • The surroundings regain visible descriptors knowledge employing the RFID code and transmit the details to the machine using the uncorrected transmission. • The identification procedure has three phases - colors-based filtering, blob filtering, and LHD-based matching.	• The experiment is set up at the Korea Institute of Industrial Technology. • The object identification network has MPEG-7 visual descriptors and detector networks.

4.4.5 RFID-enabled factories/logistics

RFID logistics stocks and administers the records based on the period series. It manages the complicated reasoning association among massive datasets.

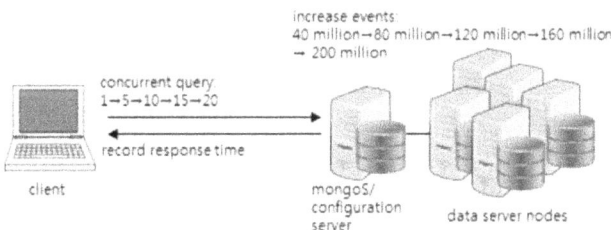

Figure 4.7 test scenario (Kang, Park, & Youm, 2016).

MongoDB cluster (Kang, Park, & Youm, 2016) has six hubs with a similar configuration. Five of them are knowledge-keeping devices. The last one has a MongoS and specification gateway. The learning-questioning buyer procedure is in the identical web as MongoS. It tests in connected surroundings. Figure 4.7 represents the same. Table 4.3 summarizes the Contribution towards RFID-Logistics

Table 4.3 Contribution towards RFID-Logistics

Contribution	Description of work	Experiment specification
(Qiu, 2007)	• An international label code recognizes the component. • It has two subcomponents. It has the distinctive title of the labelled creation and the Internet Protocol (IP) address. • The system allocates regulations to all the pieces. • The system provides relevant data to the customer.	• A test-bed design has ten parts of the mechanical packaging and trial tools to authenticate the suggested architecture. • Operations contain entrance support, die-attach, cure, wire bonding, plasma maintenance, review, abnormality, and departure pattern. • The architecture employs eight Internet RFID readers and 60 markers in the performed trials.
(Hameed, Khan, Dürr, & Rothermel, 2010)	• The production sequences are production approaches. • The industrial place splits into topological provinces. • The territory is related to the virtual reader. It assembles unprocessed information from a collection of physical RFID readers in its location. • The system employs data to infer arrangements of entities carrying using the production tracks.	• The simulations execute with 8,192 physical readers. • It has 1024 nodes.
(Feng, Li, Xu, & Zhong, 2018)	• The system transmits the cleansed data to assemble. • It extracts data from data stored in the warehouses.	• The system uses 413 472 pieces of data. • The UserID has 375 386 information. • It results in 1353 identical data points.
(Lee & Chan, 2009)	• The plants provide finished yields to suppliers to disseminate them to clients. • RFID studies the amount of repaid goods and recyclable elements. • A genetic algorithm is a meta-heuristics method used to decide the best place of gathering points by enhancing the portion of reciprocated goods.	• population size is 20 • coverage distance is 20 km • total monthly return is 927 units • the proposal uses IBM computer with 3.19 GHz CPU and 512 MB RAM

4.5 Future scope

The destiny of RFID is evolving and growing as more enterprises and firms finance the trend. RFID labels connect with detectors, thin-film photovoltaic solar cells, and other appliances. The setup can aid the businesses to watch and control investments and freight. Passive detectors for climate, wetness, stress, and deliver intellect. RFID allows the introduction of appliances in trade, healthcare, production, and other domains. The corporations provide resolutions without the classic approval and deployment prices with the help of cloud infrastructure.

References

Ahuja, S., & Potti, P. (2010). An introduction to RFID technology. *Commun. Netw, 2*(3), 183-186.

Al-Ali, A. R., Aloul, F. A., Aji, N. R., Al-Zarouni, A. A., & Fakhro, N. H. (2008). Mobile RFID tracking system. *3rd international conference on information and communication technologies: from theory to applications* (pp. 1-4). Damascus, Syria: IEEE.

Amendola, S., Lodato, R., Manzari, S., Occhiuzzi, C., & Marrocco, G. (2014). RFID technology for IoT-based personal healthcare in smart spaces. *IEEE Internet of things journal, 1*(2), 144-152.

Baeg, S. H., Park, J. H., Koh, J., Park, K. W., & Baeg, M. H. (2007). Building a smart home environment for service robots based on RFID and sensor networks. . *International Conference on Control, Automation and Systems* (pp. 1078-1082). Seoul, Korea (South): IEEE.

Bouchard, K., Fortin-Simard, D., Gaboury, S., Bouchard, B., & Bouzouane, A. (2013). Accurate RFID trilateration to learn and recognize spatial activities in smart environment. *International Journal of Distributed Sensor Networks, 9*(6), 936816.

Chen, J. L., Chen, M. C., Chen, C. W., & Chang, Y. C. (2007). Architecture design and performance evaluation of RFID object tracking systems. *Computer Communications, 30*(9), 2070-2086.

Escobedo, P., Carvajal, M., Capitán-Vallvey, L., Fernández-Salmerón, J., Martínez-Olmos, A., & Palma, A. (2016). Passive UHF RFID Tag for Multispectral Assessment. *Sensors, 16*, 1085.

Faschinger, M., Sastry, C. R., Patel, A. H., & Tas, N. C. (2007). An RFID and wireless sensor network-based implementation of workflow optimization. *International Symposium on a World of Wireless, Mobile and Multimedia Networks* (pp. 1-8). Espoo, Finland: IEEE.

Feng, J., Li, F., Xu, C., & Zhong, R. Y. (2018). Data-driven analysis for RFID-enabled smart factory: A case study. *IEEE Transactions on Systems, Man, and Cybernetics: Systems, 50*(1), 81-88.

Figueroa, S., Añorga, J., & Arrizabalaga, S. (2019). An attribute-based access control model in RFID systems based on blockchain decentralized applications for healthcare environments. *Computers, 8*(3), 57.

Golding, P., & Tennant, V. (2008). Evaluation of a radio frequency identification (RFID) library system: preliminary results. *International Journal of Multimedia and Ubiquitous Engineering, 3*(1), 1-18.

Hameed, B., Khan, I., Dürr, F., & Rothermel, K. (2010). An RFID based consistency management framework for production monitoring in a smart real-time factory. *Internet of Things (IOT)* (pp. 1-8). Tokyo, Japan: IEEE.

Huang, S., Guo, Y., Zha, S., Wang, F., & Fang, W. (2017). A real-time location system based on RFID and UWB for digital manufacturing workshop. *Manufacturing Systems 4.0 – Proceedings of the 50th CIRP Conference on Manufacturing Systems. 63*, pp. 132-137. Taichung City, Taiwan.: ELSEVIER.

Jaykumar, J., & Blessy, A. (2014). Secure smart environment using IOT based on RFID. *International Journal of Computer Science and Information Technologies, 5*(2), 2493-2496.

Joshi, G. P., Acharya, S., Kim, C. S., Kim, B. S., & Kim, S. W. (2014). Smart solutions in elderly care facilities with RFID system and its integration with wireless sensor networks. *International Journal of Distributed Sensor Networks, 10*(8), 1-11.

Kang, Y. S., Park, I. H., & Youm, S. (2016). Performance prediction of a MongoDB-based traceability system in smart factory supply chains. *Sensors, 16*(12), 2126.

Lee, C. K., & Chan, T. M. (2009). Development of RFID-based reverse logistics system. *Expert Systems with Applications, 36*(5), 9299-9307.

Mora-Mora, H., Gilart-Iglesias, V., Gil, D., & Sirvent-Llamas, A. (2015). A computational architecture based on RFID sensors for traceability in smart cities. *Sensors, 15*(6), 13591-13626.

Park, C. R., & Eom, K. H. (2011). RFID label tag design for metallic surface environments. *Sensors, 11*(1), 938-948.

Park, J. H., Baeg, S. H., Koh, J., Park, K. W., & Baeg, M. H. (2007). A new object recognition system for service robots in the smart environment. *International Conference on Control, Automation and Systems* (pp. 1083-1087). Seoul, Korea (South): IEEE.

Peng, S. L., Liu, C. J., He, J., Yu, H. N., & Li, F. (2019). Optimization RFID-enabled Retail Store Management with Complex Event Processing. *International Journal of Automation and Computing, 16*(1), 52-64.

Pillai, V., Heinrich, H., Dieska, D., Nikitin, P. V., Martinez, R., & Rao, K. S. (2007). An ultra-low-power long range battery/passive

RFID tag for UHF and microwave bands with a current consumption of 700 nA at 1.5 V. *IEEE Transactions on Circuits and Systems I: Regular Papers, 54*(7), 1500-1512.

Poigai Arunachalam, S., Sir, M. M., Sadosty, A., Nestler, D., Hellmich, T., & Pasupathy, K. S. (2017). Optimizing emergency department workflow using radio frequency identification device (RFID) data analytics. *Frontiers in Biomedical Devices, 40672*, 1-2.

Qing, X., & Chen, Z. N. (2007). Proximity effects of metallic environments on high frequency RFID reader antenna: Study and applications. *IEEE transactions on Antennas and Propagation, 55*(11), 3105-3111.

Qiu, R. G. (2007). RFID-enabled automation in support of factory integration. *Robotics and Computer-Integrated Manufacturing, 23*(6), 677-683.

Sunny, A. I., Tian, G. Y., Zhang, J., & Pal, M. (2016). Low frequency (LF) RFID sensors and selective transient feature extraction for corrosion characterisation. *Sensors and Actuators A: Physical, 241*, 34-43.

Wang, C., Shi, Z., & Wu, F. (2017). An improved particle swarm optimization-based feed-forward neural network combined with RFID sensors to indoor localization. *Information, 8*(1), 9.

Want, R. (2006). An introduction to RFID technology. *IEEE pervasive computing, 5*(1), 25-33.

Xie, H., Shi, W., & Issa, R. R. (2011). Using rfid and real-time virtual reality simulation for optimization in steel construction. *Journal of Information Technology in Construction (ITcon), 16*(19), 291-308.

Yang, L., Rida, A., Vyas, R., & Tentzeris, M. M. (2007). RFID tag and RF structures on a paper substrate using inkjet-printing technology. *IEEE transactions on microwave theory and techniques, 55*(12), 2894-2901.

Zhang, J., Huang, B., Zhang, G., & Tian, G. Y. (2018). Wireless passive ultra high frequency RFID antenna sensor for surface crack monitoring and quantitative analysis. *Sensors, 18*(7), 1-11.

Zhao, B., Kuo, N. C., & Niknejad, A. M. (2016). An inductive-coupling blocker rejection technique for miniature RFID tag. *IEEE Transactions on Circuits and Systems I: Regular Papers, 63*(8), 1305-1315.

www.ingramcontent.com/pod-product-compliance
Lightning Source LLC
Chambersburg PA
CBHW050501190326
41458CB00005B/1383